Sex, Drugs, and

THE OCEANS'
ODDEST CREATURES
AND WHY THEY
MATTER

SEX,
DRUGS,
AND
Sea Slime

ELLEN PRAGER

The University of Chicago Press   Chicago and London

DR. ELLEN PRAGER is a marine scientist and author, formerly the chief scientist at *Aquarius Reef Base* in Key Largo, Florida, which hosts the world's only operating undersea research station, and at one time assistant dean at the University of Miami's Rosenstiel School of Marine and Atmospheric Science. Dr. Prager now focuses much of her efforts on bringing ocean and earth science to the public through books, public speaking, and working with the media. She has appeared on the *Today Show*, *Good Morning America*, CNN, *Larry King*, and more. Her publications include *Chasing Science at Sea*, *The Oceans*, and *Furious Earth: The Science and Nature of Earthquakes, Volcanoes, and Tsunamis*. Her children's books include *Sand*, *Volcano*, and *Earthquakes* with the National Geographic Society, and a novel, *Adventure on Dolphin Island*.

The University of Chicago Press, Chicago 60637
The University of Chicago Press, Ltd., London
© 2011 by Ellen Prager
All rights reserved. Published 2011.
Printed in the United States of America

20 19 18 17 16 15 14 13 12 11   1 2 3 4 5

ISBN-13: 978-0-226-67872-6 (cloth)
ISBN-10: 0-226-67872-5 (cloth)

Library of Congress Cataloging-in-Publication Data

Prager, Ellen J.
    Sex, drugs, and sea slime : the oceans' oddest creatures and why they matter / Ellen Prager.
        p. cm.
    Includes bibliographical references.
    ISBN-13: 978-0-226-67872-6   (cloth : alk. paper)
    ISBN-10: 0-226-67872-5   (cloth : alk. paper)
    1. Marine organisms. 2. Marine animals. 3. Marine pharmacology.
4. Sexual behavior in animals. 5. Zoology, Economic.   I. Title.
    QL122.P73 2011
    591.77—dc22

                                                    2010037850

*To my colleagues, who have
dedicated long hours and hard work
to the study of the oceans and marine
life, and whose efforts are unveiling
the mysteries of the sea and helping
us to understand their connections
to everyday life.*

CONTENTS

A NOTE ON THE TITLE

Secreted away in stacks of scientific journals and academic books, hidden within the digital cloud of data on the web, is a wealth of fascinating information about marine organisms and their odd strategies for living in the sea. Years of research by scientists across the world have produced some extraordinary, almost unbelievable, stories about the oceans' residents, many of which have surprising connections to society and everyday life. This information is all too rarely revealed to the layperson or made accessible and engaging to the nonscientist. Yet much of what has been discovered contributes significantly to our understanding of the planet, the oceans, and our reliance on marine resources, and is at the same time wonderfully entertaining, much of it being stranger than fiction and more salacious than a soap opera.

This book is meant to reveal the true goings-on in the sea, a tell-all of the oceans. It is not meant to be a comprehensive biological text or an exacting overview of the problems faced in the oceans, but rather a brief and entertaining look at some of the oceans' most fascinating creatures, their unusual tactics for survival, and their invaluable links to humankind. The end goal is to showcase the importance of the great diversity of life in the sea, why it is at risk, and why we all should care.

The original title for the book was going to be something like *Weird and Wild Under the Sea: And Why These Creatures Matter*. As I began to comb through the literature and talk to my colleagues, however, the title began to evolve as several somewhat surprising and rather captivating themes started to emerge. I discovered that more of the oceans' residents use, deploy, or are made up of slime than I ever expected. They use mucus to capture food, defend against predators, clean themselves off, and in reproduction. Many organisms have gelatinous bodies or arms of goo, and some use slime as protection against the cold, to reduce drag, or to enhance their travels by slickening their path. The undersea world is a seriously slimy place.

While I knew that the citizens of the sea tend to have three main goals in life, to find or produce food, to avoid being eaten, and to reproduce, I never realized the extent of the strange sex going on in the oceans. The instinct undersea to procreate is strong, and creatures have evolved many intriguing strategies to attract mates, copulate, and improve the likelihood that their progeny will be born and survive. Group sex and partner switching are common, while monogamy and fidelity are rare. Some marine organisms can be both sexes at once, while others change gender when the need arises—they are transgender on call. In some species, the difference between the sexes is extreme and mating habits are somewhat unusual, to say the least. While coupling, organisms in the oceans have to contend with changing conditions, sinking, and predators. Sometimes, even in the middle of the action, their own partners are a threat. Marine organisms are often outfitted with accoutrements designed specifically for sex, and some are especially well-endowed. Below the waves, cloning, seduction, and competition among rivals are rampant. And some creatures must literally give it their all in mating or in the incubation of their offspring, as breeding is a precursor to life's end. The oceans are a den of strange sexual relations that enables life to exist and genes to pass from one generation to the next.

As I began investigating the connections between society and marine life, another topic of commonality emerged. The oceans' organisms have long provided an invaluable source of food, jobs, and economic revenue, but increasingly they are also supplying new drugs and being used as models for biomedical research and in biotechnology. Already, basing their work on compounds derived from marine organisms, researchers have developed effective medicines to combat cancer and HIV, and to treat pain and inflammation. Nobel Prizes have been won based on studies that used marine organisms as biomedical models, and the field is widening and producing new and promising results. Even after years of hearing and reading about my colleagues' endeavors in this area, I was surprised by the increasing number and diversity of marine organisms that are being studied to improve human health. A revision of the title seemed warranted, and the book became *Sex, Drugs, and Sea Slime: The Oceans' Oddest Creatures and Why They Matter.*

A few brief points before diving in for a behind-the-scenes look at the heart of the sea, the lifeblood of the oceans—the creatures. Our understanding of marine life is superficial, for the most part. We don't know how many organisms there are or what most do on a daily basis. Over the last decade, the Census of Marine Life program has worked to increase our inventory of the species in the oceans, especially in hard-to-access environments, such as in the deep sea, in polar waters, and on isolated seamounts. Researchers from across the globe successfully documented tens of thousands of organisms, thousands of which are new to science. Yet we know that there are many more creatures that have yet to be discovered. In the pages of this book, I have attempted to compile some of the most interesting information about marine organisms gleaned from years of research. The behaviors described are based on observations in the wild or in captivity or extrapolated from the stomach contents, molecular biology, or anatomy and physiology of specimens collected. Sometimes, the information is the best guess of the experts based on the data at hand. As scientists gather additional information, our understanding of these creatures will improve and be refined. And unquestionably new surprises will arise.

Throughout the book the traditional divisions of the oceans into shallow or surface waters, the midwater or twilight zone, and the deep sea are used. Shallow water typically means from the surface down to a depth of about 200 meters (660 feet). The midwater or twilight region is from 200 to about 1,000 meters (660 to 3,300 feet) and the deep sea is below that. In addition, in discussions of the distribution of organisms, the term *Indo-Pacific* is frequently used. This region, comprised of the Indian Ocean, western and central Pacific Ocean, and the seas around Indonesia, is considered a hotbed of diversity within the oceans and contains many of the sea's most interesting and unusual creatures.

The oceans are incredibly rich with life, and in this brief text I could not include all of the sea's remarkable creatures or everyone's favorites. Maybe there will have to be a sequel. I hope that you will be amazed and fascinated and sometimes laugh at the organisms that are described herein and their peculiar ways of living in the sea. And that you will come away with a new appreciation for how marine life and the oceans are relevant to humankind and the great risks we face if species are lost. Enjoy, chuckle, and be amazed as you dive into the bizarre world of life on, in, and under the sea, and discover why the citizens of the oceans deserve our attention and need our help.

One last note regarding the tragic 2010 Deepwater Horizon accident and oil spill in the Gulf of Mexico. My heart goes out to the families who lost loved ones and those suffering from economic and emotional distress, as well as to the innocent creatures of the Gulf and associated environments that have been killed or suffered due to this preposterous blunder by human beings. The blame for the accident is not only on British Petroleum or Transocean, though they are clearly responsible and must foot the bill; it also falls on the government agencies involved and the rest of the oil industry, all of whom were completely unprepared to handle the accident or cleanup, and allowed the negligence that enabled it to happen in the first place. The long-term impact on the Gulf of Mexico and surrounding coastlines is unknown. The oceans are resilient and the Gulf will come back, but it may take years and never be quite the same. We can only hope that the lessons learned from this disaster will prevent another such accident from ever occurring, and that they will make us all more aware of the links between a healthy ocean and a vibrant, economically stable society.

1 The Invisible Crowd

For many people the oceans inspire a sense of peace, a feeling of serenity. Yet even in the calmest of waters, in the bluest of blue, there is a crowd jostling about and a timeless battle being waged. When you swim in the sea, they surround you. When you look into the oceans' depths, they are there unseen, right before your eyes. Much of the sea is actually a crowded place, swarming with organisms struggling to survive and aspiring to reproduce. The majority of these creatures are small, too small to see, including tiny plants, animals, bacteria, and other microbes. In just one teaspoon of seawater there may be millions of organisms. Drinking seawater—not a good idea for a whole host of reasons! The oceans' invisible realm is packed with a strange menagerie of life forms whose diversity comes from the struggles of living in the sea's zone of small. They must find ways to stay afloat and obtain the necessary resources to grow and to reproduce, all the while staying in favorable conditions and if possible, avoiding becoming someone else's lunch. Within the oceans' invisible crowd there is great beauty along with some serious slime and a monster or two, albeit in miniature.

SMALL DESIGNS

For the oceans' tiny floating plants, the phytoplankton, wants are relatively simple: they need sunlight, nutrients, carbon dioxide, and favorable temperatures. These, the needs for photosynthesis, seem like uncomplicated desires, but satisfying them is not always so easy in the oceans. Phytoplankton tend to sink, and most of the oceans' living space is dark, well below the reach of sunlight. The small creatures of the sea must also deal with the oceans' incessantly changing patterns of flow produced by currents, circular eddies, waves, upwellings, downwellings, and the wind. The concentrations of nutrients in the sea, which phytoplankton need as fertilizer to grow, also vary strongly over depth, in time, and geographically.

Over millions of years, phytoplankton have evolved a remarkable array of beautiful and odd forms that allow them to satisfy their needs and to flourish. Three of the most common and abundant phytoplankton are illustrative of the beauty and strange diversity within their diminutive ranks; these are the delicate diatoms, the spaceship-resembling coccolithophores, and the multitalented dinoflagellates.

Diatoms are small, single-celled algae with a thin, glassy *test*, or shell, made of silica. They are akin to miniature greenhouses with tiny glass walls surrounding specks of golden-green plant material. In today's oceans, there are believed to be thousands of diatom species with a wide variety of architectural forms. They may form disks or cylinders, or take on the shape of a square, triangle, or oblong pendant. To reduce sinking, diatoms often link up (not to be confused with the popular term "hooking up") to create delicate linear, circular, or twisting chains. Under a microscope, such chains resemble elegant necklaces of spun glass bejeweled with light-colored emeralds. Pits, perforations, spines, or spikes add ornamentation to the diatoms and also help to reduce sinking and facilitate more efficient exchange of nutrients and wastes. In addition to leading a life afloat, diatoms may dwell at the seafloor and can grow on just about any surface within the shallow sunlit waters of the sea. The slimy black goo that makes a dock slippery, grows on your boat, or adorns the belly of a whale is in all likelihood a lush coating of diatoms.

The diatoms have evolved an especially efficient means to reproduce. They multiply through simple asexual division in which the top and bottom parts of the diatom split apart and each produces a smaller new half. The only problem with this quick-fire method of reproduction is that over time and with successive generations, diatoms get smaller and smaller, potentially replicating themselves into extinction. Eventually, a tiny diatom must cast off both portions of its silica test and form a special type of cell that can produce a larger-sized diatom. When temperature, nutrients, and light are all favorable for growth, diatoms can reproduce rapidly and in great abundance. In less than three weeks just one small diatom can produce as many as one million daughter cells.

Coccolithophores are the oceans' alien spacecrafts in miniature, with a really big one growing to some 40 microns in size, about half the width of a thick piece of human hair. They are unicellular brown algae that have a spherical hull or covering of overlapping, microscopic plates, called coccoliths. Each coccolith is made of calcium carbonate (limestone) and bears a striking resemblance to an ornately designed hubcap. Among the different coccolithophore species, there is a wealth of hubcap designs, and each individual may be covered with as many as one hundred of them, though the average appears to be about ten to twenty. Some coccolithophores constantly shed their coccoliths and then produce replacements, while others make them only if a repair is needed. It is unclear what purpose the coccoliths serve. One theory suggests that the calcite sheath they create offers protection from high-intensity sunlight, allowing the coccolithophores to flourish at the surface in the tropics and subtropics. Another thought is that the coccoliths actually concentrate light toward the organisms' interior so that they can survive in low light conditions. Then again, others hypothesize that the coccolithophores' hubcaps provide physical protection or act as floatation devices.

Like diatoms, coccolithophores can reproduce in great abundance under favorable conditions. Images from space have documented huge, milky white blooms of coccolithophores hundreds of miles across, encompassing an area as large as England. The famed white cliffs of Dover are a massive accumulation of coccoliths, evidence of a once enormous population of coccolithophores. Many of the coccoliths that get deposited on the seafloor and create such accumulations have probably been transported via special delivery—packed in a protective coating of poop-produced slime. In the oceans, many organisms release their wastes in gooey pellets, which aggregate fine particles into bigger, slime-encased packets. Because they sink faster than smaller bits of waste, the oceans' balls of poo are less likely to be consumed, drift away, or dissolve before reaching the seabed.

Dinoflagellates are also small and single-celled; they are the oceans' version of miniature whirling dervishes. Two tiny, whip-like tails, or flagella, cause them to whirl and spin when they swim. Dinoflagellates are a diverse and versatile bunch, with amazing, and in some cases harmful, attributes. One common form resembles a miniature grappling hook, while others are tooth-shaped or take on the appearance of a spinning top. Some dinoflagellates have an outer coating of chitin (a very thin version of a crab or shrimp's shell), while others go naked. They may sport spines, wings, or even horns. Dinoflagellates often contain chlorophyll for photosynthesis, but some can also engulf food particles like an amoeba—they are switch-hitters, able to behave like plants or animals depending on the situation. Many dinoflagellates are stowaways within other organisms, such as corals, living in a mutually beneficial or symbiotic relationship with their hosts. Others of these tiny organisms harbor harmful toxins or can form cysts and lie temporarily dormant on the seafloor—just waiting for their time to prosper. Dinoflagellates are also responsible for much of the oceans' remarkable nighttime bioluminescence; producing the bright twinkling of light in a ship's wake or the sparkling trail that traces a diver's movements. Given the right temperature, light, and nutrient conditions, like diatoms and coccolithophores, the dinoflagellates can bloom, reproducing quickly and in great number. A thousand or so species of these superadaptable organisms are recognized today, but undoubtedly there are many more yet to be identified.

Another member of the phytoplankton clan that deserves at least a quick mention are the cyanobacteria, some of the oldest organisms on Earth, believed to have originated about three billion years ago. Cyanobacteria are often abundant in nutrient-poor areas where other phytoplankton are scarce. The cyanobacteria *trichodesmium* is common in warm, tropical waters and is often mistaken for little bits of pollution or floating pieces of general ocean "schmutz." It forms small floating chains that look like short needles, tufts or little tan puffs. *Trichodesmium* can also aggregate into larger mats that provide an important habitat for small fishes and other organisms at the surface. Some of the cyanobacteria have the ability to do something that is rare in the oceans: they can "fix" nitrogen, or transform it from a free molecular state into a more plant-friendly form. There is growing recognition that the cyanobacteria probably play a more important role in the sea than previously thought.

The zooplankton are the sea's floating or weakly swimming animals and are more diverse than its host of tiny plants, with an even greater array of small designs. They populate the world's oceans in abundance, from the deepest of deep-sea trenches to the shallowest tide pool. Within this varied group of creatures there are some strange goings-on: slime-wielding grazers, gluttonous predators that are small in scale, but huge in appetite, and even a few cross-dressers that can photosynthesize like plants, but when needed, switch sides and capture prey as animals. Within the sea, zooplankton are just about everywhere and they make up a large proportion of what goes on, unseen, in seawater.

ARMS OF GOO

Two of the most common and beautiful members of the zooplankton are the small foraminifera and radiolarians. The foraminifera are essentially amoebas that live within a shell of calcium carbonate, typically the size of a sand grain. Their limestone homes may be a single small orb or a cluster of them, or form a miniature nautilus-like encasing. As foraminifera grow, they add additional rooms to their homes by producing new chambers in their shells. Some have long, calcite spines, and many have pits and perforations; they have been likened to tiny, pitted potatoes. Except this potato has arms of goo that it can extend out of its shell to use as a slimy net to capture smaller organisms, such as bacteria and phytoplankton. Some foraminifera also have dinoflagellates living within their tissues that aid in growth through photosynthesis. These foraminifera are much like sophisticated satellites that open their solar panels at daybreak. In the morning, the foraminifera transport the algae out of their shells to the ends of their spines and gooey arms (plate 1). There, the dinoflagellates capture sunlight and grow through photosynthesis; in the process they use up the foraminiferas' wastes while producing oxygen. As darkness falls, the algae are drawn back inside the foraminiferas' shells. Thousands of symbiotic algae may be housed within a foraminifera's tissues, making it a floating, dense packet of productivity. There are some forty species of planktonic foraminifera, with additional varieties living at the seafloor.

Radiolarians are also small, mainly spherical, creatures, but their shells are internal (skeletons), are made of silica, and tend to be highly decorated with spines. Their skeletal architecture is also more open than the shells of their calcareous relatives, like the structural framework of a minuscule building in the round. The radiolarians have arms of goo as well, which are used to dine on prey captured within a foamy net of slime that surrounds their shells and spines. In the tropics, radiolarians can also form unusually large colonies that may grow to be meters in length. Where radiolarians and foraminifera are abundant, a carpet of tiny shells may accumulate on the seabed, creating soft sediments known as siliceous or calcareous oozes, respectively.

AN EPIC BATTLE

The foraminifera and radiolarians float in relative peace about the sea, ensnaring their food in arms of goo and sometimes succumbing to the whims of nature as they are captured and ingested by other organisms. All around them, however, an epic but invisible battle is being waged. It is a struggle for survival, a battle for supremacy. Okay, maybe not quite that dramatic, in fact, it is more like Godzilla versus a cow. The combatants are two of the most common and abundant organisms in the sea, and the weaponry unleashed, at least by one of them, is formidable.

The opponents in this everlasting war are the copepods and arrow worms. Copepods are small, shrimp-like crustaceans that feed mainly on hapless bacteria, dinoflagellates, and other phytoplankton. They are simple grazers, foraging and filtering the oceans' waters. On the other side of the ring is the arrow worm, also known as a chaetognath (pronounced "*kee*-tahg-nath"). Though diminutive in size, the arrow worm is one of the sea's most voracious and lethal carnivores, driven by a constant hunger and equipped with an array of deadly armaments. It is clearly an unfair fight that leaves the copepod living a perilous life as the eternal underdog to the arrow worm.

What the copepods lack in weaponry, they make up for in sheer abundance. Copepods are considered the most numerous of multicelled organisms in water, and surprisingly, probably the largest source of protein in the sea (it is also surprising that we have not found some way to make them palatable for human consumption). They are found in all marine environments, from the surface to the oceans' depths, under the ice, and at deep-sea hydrothermal vents. There are thousands of copepod species, and when you see small dots flitting about in the sea or in an aquarium—a sure bet is copepods. Individuals are typically small, averaging just a few millimeters in length. Like other crustaceans, they have an external carapace or shell made of chitin and numerous jointed limbs. The copepod's body is short, segmented, and elliptical, and each has a single eye in the middle of its head—one freshwater species is appropriately named cyclops. Its two characteristic antennae may be long and delicate or short and stubby. Copepods also have oar-like swimming legs, which are used to move about, typically at a slow and steady pace. If, however, some speed is needed, copepods can swim at a faster clip through a series of jerky bursts. Copepods also use their swimming appendages to create currents that circulate water around their bodies and bring food particles into reach. While most of the oceans' copepods dine on small particles and phytoplankton, some have cannibalistic predilections and will feast on other copepods. There are also a few species that dine on pieces of flesh torn from unwary fishes or that live as parasites feeding on skin or blood. Though they are adaptive creatures and can produce quick bursts of speed, copepods are still, simply no match for the ever-voracious arrow worm.

In the oceans' domain of the unseen, the arrow worm is the top dog, the ultimate predator. It is built exceptionally well for the life of a microscopic hunter. Most are transparent, providing for excellent camouflage and stealth in the open sea. They are also torpedo-shaped for speed with lateral fins for stability and have relatively large paddles as tails that provide for efficient propulsion. To seek out their prey, arrow worms use a speedy darting or hop-and-sink pattern of swimming. Although they have two eyes, vision is probably not their most important sense when in search of food, aka victims. Their entire bodies are lined by hair-like receptors that attach to nerves, which can sense nearby vibrations or water movements, making each arrow worm one big (but actually small), sensitive motion detector.

When a potential food item is detected, with just a simple flex and flick of its tail, the arrow worm goes on the attack. It reaches out with long, zinc-hardened, and hooked grasping arms that are conveniently attached to its head. Prey are grabbed and pulled toward the arrow worm's mouth, where sharp, tiny teeth puncture and immobilize it. If that doesn't do the job, a bacteria-derived venom is injected to cause paralysis. It is the same nerve toxin used by the deadly blue-ringed octopus. The arrow worm then projects its mouth like a snake over the ill-fated prey, which is literally stuffed down its throat and into the gut for digestion. Mucus glands provide slimy lubrication and allow for easier swallowing. If the prey is too big, the arrow worm simply folds it over, as a person eating a slice of pizza might. And arrow worms can be gluttonous creatures, giving a whole new meaning to the concept of overeating—sometimes they attempt to consume food items that are just a little too big, causing them to choke to death. Not the smartest creature on the block, but then again they don't actually have a brain. They are also not above eating their own—cannibalism is common. And if an arrow worm is really hungry, one is just not enough, and it will consume several victims in rapid succession—with the possibility once again of death by overeating. In addition to copepods, arrow worms like to prey on barnacles and fish larvae. Fortunately, these dangerous predators range in size from a few millimeters up to merely several centimeters (less than an inch) in length. Zooplankton expert Peter Ortner at the University of Miami's Rosenstiel School of Marine and Atmospheric Science suspects that the arrow worm was the model that inspired the vicious sand worms in Frank Herbert's classic *Dune*.

PACKETS OF PROTEIN

Among the zooplankton, there are many odd shrimp-like crustaceans. In fact there are hundreds of species that seem similar to copepods but are distinctly different, including the mysids, amphipods, isopods, and the famous euphausiids. The euphausiids are better known as krill, and they are little packets of protein, or maybe the oceans' equivalent of an energy drink. In the Antarctic, krill, especially the species *Euphausia superba*, are found not just in great abundance, but in aggregations so huge they are hard to imagine. Individual krill are just a few centimeters in length, but when they get together en masse they can form superswarms that span an area over one kilometer across and several hundred meters deep. Within one superswarm there may be as many as sixty thousand krill per cubic meter. Because 50 to 60 percent of a euphausiid's body weight is protein, these aggregations represent a whole lot of energy packed into a small amount of ocean. Swarming may be advantageous for krill in foraging and as a means to reduce the likelihood of individuals being eaten by predators.

For the other creatures of the sea, these often long-lasting, packed-with-protein aggregations of krill are an energy-laden feast. They attract a wide variety of diners, especially baleen whales, such as the humpback, blue, and southern right whales, along with leopard and elephant seals, penguins, and fishes. Krill themselves are opportunistic omnivores; that is they are adaptive feeders. When phytoplankton are plentiful, they gorge on the sea's floating plants, but when veggies are scarce, they will consume zooplankton, scavenge for detritus at the seafloor, or even pick algae off the undersides of the ice. Krill may also be able to go for relatively long periods of time without food, though they shrink in size when fasting. Their abundance is aided by the fact that a single female euphausiid can lay up to ten thousand eggs at a time, and may spawn several times during a breeding season. One estimate suggests that there may be as much as 500 million tons of krill in the Antarctic—it is an amazing abundance that is critical to the diversity of life that resides within the oceans' most southern reaches.

MYSTERY YOUNG

Many of the strangest members of the zooplankton are only temporary visitors within the realm of the invisible. These creatures have planktonic larvae, but as adults they live at the seafloor or are among the oceans' full-fledged swimmers. They are often the oceans' mystery young; animals that in their infancy look little like their fully-grown versions. An excellent and odd example is the lobster. The adult, of course, has a familiar form with a hardened carapace, waving antennae, two large claws or spines at the front, lots of jointed, spindly legs, and a flattened tail. The lobster larva, however, is not so recognizable. It resembles a white, paper-thin, Halloween skeleton or a squashed, leggy spider rather than the crustacean we have come to know, love, and dine on. The larval sea star is another trickster, resembling an octopus, with seemingly too many arms jutting out from a transparent blob of a body. At a young age, the identity of worms, barnacles, and sea cucumbers is also elusive. When small, baby snails and other mollusks have a retractable, wing-like sheet of tissue that helps them to stay afloat, making them a miniature-in-flight form of the adults. Tiny jellyfish, baby squids, and larval fishes look more like their grown-up forms, but distinguishing species at an early age can be near to impossible. The baby crab, called a zoea, is arguably one of the cutest of the floating young, a Pac-Man sort of creature. It has relatively large eyes, a small body shaped like a pith helmet with a dorsal spine, and a long rostrum or beak. The zoea is not, however, the crab's only awkward phase in its youth; it will grow and molt through several changes and odd appearances before attaining a fully developed, more familiar form.

BIGGER, BUT INVISIBLE

Within the oceans' unseen and floating club, there are also a few larger but still transparent members. Included within this group are jellyfish, along with their look-alikes, the siphonophores, as well as the comb jellies and salps. Comb jellies or ctenophores (pronounced "*teen-o-fours*") are not true jellyfish and lack the stinging cells of their transparent cohorts. Among their kind are the sea gooseberries and walnut varieties, whose names are indicative of their typical size and shape. Comb jellies are essentially see-through balls of goo that move about the ocean using eight rows of combs that line their bodies vertically and are made up of thousands of hair-like cilia. As if repeatedly doing "the wave," the cilia beat to propel the comb jellies smoothly through the water. When sunlight strikes their beating cilia, they become bejeweled, decorated by beautiful rainbows of iridescent and flickering stripes. Comb jellies are also surprisingly voracious predators, with some using sticky tentacles to ensnare copepods, larvae, and other plankton. Other species have extended lobes around their mouths and mucus that are used to entrap food particles. One type of comb jelly is notoriously cannibalistic, preferentially dining on its own kind. The *Beroe* species use their large mouths to engulf their brethren and then swallow them whole. They will even attempt to consume comb jellies that are nearly their same size, quite the mouthful. Comb jellies are found in a wide range of marine environments, mainly in relatively shallow water, but a few species can live as deep as about 3,000 meters (9,800 feet). Reproduction is fairly simple and efficient, with each comb jelly playing the roles of both male and female by releasing eggs and sperm into the water, where fertilization occurs and larvae develop.

For salps, reproduction is a bit more complicated, and in fact, their baby-making ways are quite unusual; they produce hundreds of clones-on-a-string and then breed sexually. Salps are simple, barrel-shaped, transparent creatures up to about 10 centimeters (4 inches) long. They are propelled through the water by contracting bands of muscle that encircle their bodies, forcing water in one end and out the other. As water passes through their bodies, nets of mucus inside strain out zooplankton and particulate material for food. In a sense the salps are excellent undersea multitaskers, since their swimming mechanism is also how they feed. They are also prodigious poopers. They can filter large quantities of water, liters per hour, as they swim, so they must constantly get rid of their wastes, producing copious fecal material in large, consolidated pellets.

As for their rather remarkable reproductive ways, salps live as either asexual singles or in a sexually reproducing colony of clones. The solitary salp produces its progeny by asexual budding, which generates a string of cells that divides into a chain of babies aka clones-on-a-string. The string gets longer as hundreds of young self-replicates grow, until eventually the chain breaks free of the parent and becomes the salp's colonial form. These colonies or aggregates can contain hundreds of salps in a linked chain that stretches for meters, coils like a snake, or resembles a spiraling chandelier made up of tubular crystals lying side by side. Each clone within the chain is a female carrying an egg, which is essentially a ball of jelly with a placenta that is connected to and nourished by the mother. Larry Madin of the Woods Hole Oceanographic Institution is one of the world's foremost experts on gelatinous marine life. He finds this part of the salps' reproduction to have an odd and surprising parallel to that of humans, especially given that when grown, unlike humans, the adult creature will essentially be just a muscle-lined tube of goo. After fertilization and a gestation period of about a week, a new salp is born from each female in the chain. This baby becomes the solitary form of the salp, which will later asexually bud to produce another colony of clones. After the chain of mothers gives birth, it becomes a chain of fathers. The aggregate goes male and will provide sperm to fertilize younger chains of females.

The salps' strange alternating reproductive process is very efficient, and they can reproduce quickly and in great abundance. In the open ocean, huge salp swarms may contain billions of individuals, and in the Antarctic, salps can be one of the most abundant organisms present. Madin suggests that they "are one of the most common and important animals on Earth, but are rarely seen because they live far from land, in the open ocean." His research also suggests that their prodigious production of slime-encased waste pellets results in the sinking of large quantities of organic material to the seafloor and contributes significantly to the sequestering of carbon out of the oceans' waters.

Organisms such as sea turtles, some fishes, and the jellies consume salps, while other creatures prefer to use them as a convenient condo or nursery. The devious female of one amphipod will eat its way into a salp, consume the inside tissues, and then use the carved out tube as a nursery for her young. Rumor has it that this small, gangly, and scheming crustacean was the inspiration for the ghastly creature in the film *Alien*. Other small crustaceans or fishes sometimes hang out inside a salp, using it like a condo-on-the-move. From personal experience, I can say that swimming through a thick swarm of salps is like trying to paddle your way through a vat of Jell-O, and no, there was no wrestling involved.

WHY THEY MATTER

So, why should you care about the cute crab zoea, the delicate diatom, humble copepod, ravenous arrow worm, or any of the plankton for that matter? I could write that phytoplankton are the base of the food web in most of the oceans, that over time they have produced much of the oxygen in our atmosphere, and that they act as a sink for carbon dioxide on our planet. I could stress the importance of the zooplankton as a link between the lower trophic levels in the sea (the phytoplankton) and larger marine life. For some of you this may be enough to inspire bumper stickers emblazoned with I "heart" plankton, but for many people the role of plankton in the ocean ecosystem may not seem relevant to their own lives or to society as a whole. How do phytoplankton and zooplankton relate to the food we eat, our health, the economy and jobs, or our ability to recreate and simply enjoy the beauty and fun the oceans offer? The answers might just surprise you.

Eat Fish?

Fish sandwiches, fish sticks, fish 'n' chips, fish tacos, fish fingers, sushi, sashimi, fillets, steaks, salads, and more; these are how most people think of and relate to fish. Whether at a family-friendly diner or a high-end gourmet restaurant, fish, in some form, is sure to be on the menu. And at the supermarket, fish can be found fresh, frozen, prepared, freeze-dried or in a can. Do some children now grow up thinking that fishes are naturally square and breaded?

According to the Food and Agriculture Organization of the United Nations, in 2006, Alaskan pollock, anchoveta and Japanese anchovies, Atlantic herring, chub and cero mackerel, and blue whiting were among the top ten fishes caught commercially in the oceans. These fishes all eat plankton! They are used throughout the world for human consumption and in aquaculture, and as a source of fishmeal or fish oil to feed livestock. Pollock is the fish behind many generic products, such as fish sticks or fried fish sandwiches, and it is processed to create the crab or lobster look-alike, surimi. Yellowfin and skipjack tuna are also on the top ten fisheries list, and as larvae and juveniles, they too feed on—plankton. Many of our other favorites such as salmon, lobster, and crabs feed on plankton when young or sometimes as adults. And for the fishes living in the wide sea that don't feed on plankton, the fishes they feed on probably do. So next time you bite into that fast-food fried finger of fish or enjoy a succulent bite of lobster, think plankton, because without the oceans' tiny floating plants and animals your main course would be out of stock!

In the United States we rarely eat zooplankton directly, but menus may be about to change. Krill has been called the new "pink gold." In some countries, people already partake of krill sticks or use krill as a protein filler in soups, stews, meatballs, pies, and sausages. Krill is now being marketed in the United States in assorted forms. How about a krill salad or krill and crab chowder? And that is just the beginning. Along with being a good source of protein (the whales are way ahead of us on this one), krill contain astaxanthin, which gives salmon their red color and some believe is a powerful antioxidant. Krill feed is used in twenty-six countries to feed farmed salmon, shrimp, and other fishes. Krill also contain health-boosting omega-3 fatty acids, and related products, such as lecithin, are being added to foods such as breakfast bars, cereal, and cheese. Powdered krill (one product is literally named "The Power of Krill") and krill oil are being touted as the latest in health supplements to promote everything from mental alertness and joint health to stabilized blood sugar and cardiac health, and to combat Parkinson's disease and PMS. At the time of writing, such claims remain scientifically unproven. In addition, krill oil is being used in cosmetics and other skin care products. In the future, many more of us may be eating and smearing ourselves with krill, leading some people to predict that krill will become the focus of the largest fishery on the planet.

Jobs and the Economy

Plankton are big business! Across the globe, it is estimated that over 43 million people are employed through activities directly associated with commercial fishing or aquaculture. Add to that indirectly related services such as processing, marketing, distribution, boat and equipment construction, research and development, and the number of jobs dependent on commercial fisheries escalates dramatically to an estimated 520 million. Hundreds of millions of people have jobs because of fish, and there are fish because of—plankton! Similarly, many of the top species of fish caught recreationally, including mahimahi, bluefish, Atlantic mackerel, and billfishes (marlin, sailfish, and swordfish) have larvae and juveniles that grow up eating plankton. In the United States, recreational fishing in the ocean and Great Lakes provides relaxation and enjoyment for some 10 million people each year, generating over three hundred thousand jobs along with hundreds of billions of dollars in retail, local, state, and federal tax revenues. Instead of "Show me the money," we should yell, "Show me the fish," and therefore, "Show me the plankton!"

Health

Plankton and human health are not two things we usually associate, but in truth there are numerous connections, and increasingly so, if krill becomes the next wonder food or health supplement. For now, the American Heart Association recommends that adults eat two servings of fish per week. Given that most of the fishes we consume eat plankton at some point in their lives, that means we should be eating a nice hefty serving of plankton two times per week.

Chitin and chitosan are derived from the shells of crustaceans such as krill, shrimp, and crabs. These products are used widely in society in cosmetics, shampoos, plant food, water filtration, and as antifungus agents in crops and home gardening. Based on its absorption and flocculation capabilities, chitosan is also the magic behind a new quick-healing bandage that promotes rapid blood clotting and holds great promise, especially in trauma situations or military operations.

There are also interesting links between zooplankton and disease. Some freshwater copepods feed on mosquito larvae, thus aiding in the control of malaria. On the other hand, copepods can carry the bacteria responsible for cholera. A single copepod can carry up to ten thousand cholera bacteria and may explain why outbreaks of cholera have historically followed the world's coastlines. In the estuaries of the Ganges and Brahmaputra Rivers in Bangladesh, scientists have shown that copepods can act as long-term reservoirs for cholera and may allow for periodic outbreaks.

In favorable conditions, many types of phytoplankton can bloom, reproducing rapidly and in great abundance. Unfortunately, these times of plenty can produce conditions that are deleterious to both the oceans and human health. Red tide is a type of harmful algal bloom caused by the prolific growth of dinoflagellates, some of which can harbor toxins. When released into the water, these toxins can taint shellfish, kill fishes and other marine life, and, if carried by winds onshore, cause respiratory distress in beachgoers. People who consume contaminated shellfish may experience gastrointestinal or neurologic symptoms. Even nontoxic blooms of phytoplankton can be harmful to the ocean ecosystem, and therefore to humans. In restricted or stratified waters, particularly in the summer, nutrient excess can lead to the prolific growth of phytoplankton. A massive amount of organic matter may be produced and block sunlight, thereby curtailing photosynthesis and the production of oxygen by other phytoplankton or algae. Meanwhile, bacterial decomposition of the organic matter that accumulates on the seabed uses up oxygen. Together, these processes can lead to oxygen-poor conditions and what has become known as a dead zone. Since the 1960s, and the increased use of nitrogen-based fertilizers, the number of dead zones across the globe has doubled each decade. Today there are more than four hundred ocean or coastal areas worldwide that have at one time been considered unlivable or a dead zone. Dead zones are more likely to occur where little mixing takes place and there are major population centers and watersheds nearby that deliver high concentrations of nutrients to inland seas or coastal waters. As in life, too much of a good thing can have unwanted consequences.

History Books

How do we learn about the planet's history before human habitation, without the benefit of books or archeological artifacts? What were the oceans or climate like four hundred thousand or four million years ago? Here is where our friends the plankton can help. Geologists search the world for sediments and rocks containing the preserved or fossilized shells of plankton because they house a record of what conditions were like when they were alive. The mere presence of their shells means that the overlying waters were at one time conducive to growth, with the right temperature, depth, salinity, nutrients, and light levels. Geochemical analysis of individual shells can provide additional and more detailed information. Foraminifera have proven exceptionally useful in this type of research, as have coccolithophores, radiolarians, and diatoms. Fossil dinoflagellates are increasingly being used as another means to decipher the planet's past. The plankton have been and will continue to be key to unraveling the Earth's history, especially as we try to understand climate change and its potential impacts.

Other Uses

Humans, being the ingenious species that we are, have found additional ways to use plankton. We mine ancient diatom deposits for fine-grained sediment that can be used in water filtration systems, as a mild abrasive, for insulation in boilers and blast furnaces, and as an insecticide. Have you ever visited an aquarium or owned a fishbowl? Zooplankton are an important component of the feed for aquarium life. And if you visit the Great Pyramids in Egypt, take a look at the limestone blocks used in construction. We may not know how these magnificent edifices were built, but the rocks they used clearly contain forty-million-year-old foraminifera. Their flattened disk shape suggests that they were probably a bottom-dwelling species.

Plankton may also help us to curb our reliance on fossil fuels and combat climate change. Recent research suggests that diatoms can be used to create simple and effective solar cells. The salps, as already noted, help to combat global warming by plentiful pooping of large pellets that sink to the seafloor and remove carbon from the sea. And one day, phytoplankton may provide a means of creating environment-friendly biofuels.

Nighttime Wonder

Some of the benefits that the oceans offer to humankind are hard to quantify or put a price on. The ability of the oceans to inspire wonder, creativity, contemplation, or a feeling of serenity is one of their most precious offerings, but something that is hard to assess in terms of a quantitative market value. One of the most awe-inspiring and thought-provoking sights in the oceans is bioluminescence, the biologic production of light. It may appear as an ephemeral blue-green glow emanating from the sea or a sparkling, twinkling of lights associated with objects, such as a boat, a dolphin, or a diver, passing through the water. At the sea surface at night, dinoflagellates are typically the producers of this amazing light show. They react to water movement by giving off a quick spark of blue-green light that may be their way of startling or confusing potential predators. Deeper in the sea, particularly in the oceans' twilight realm, many organisms have the capability to produce light. Some creatures produce light through a chemical reaction within their cells or special light organs called photophores; others have packets of light-producing bacteria that they can turn on, turn off, hide, or reveal, like a flashlight. Marine organisms use light in a wide variety of ways, including to communicate, to deter predators, as a burglar alarm, and to find or lure prey. Today, scientists believe that some 90 percent of the oceans' animals can produce light. Once seen, bioluminescence makes an indelible impression that evokes outright astonishment and makes inquiring minds want to know more. The ability to inspire such wonder or inquiry is not something we can put a price on or mark with a tag, but nonetheless it is of great value.

With some gooey exceptions, the members of the invisible crowd are small, but their impact on the oceans and humankind is extraordinarily large. Among the plankton there is great beauty, appetites large in scale, and organisms that congregate in megaproportion. They are the underpinning of much of the sea's miraculous web of life as well as the resources that humankind depend on. Over the past few decades, scientists have come to realize that even the smallest of small organisms in the sea, the microbes, are important. We are discovering that they are more abundant, diverse, and versatile than ever imagined, and can live in even the harshest of environments, once thought unlivable. The important role of microbes in the oceans and on the planet is just beginning to be understood. Having had the honor to sit on a panel with the famous evolutionary biologist Richard Leakey, I was intrigued when he suggested that humans may have stopped evolving and that it is now the "time of the microbes." Let us hope that we, along with all of the oceans' invisible crowd, will continue to evolve and stick around for a good while longer.

2 Mega-Slime, Seduction, and Shape-Shifting

Within the citizenry of the sea, there are some organisms whose dull or familiar countenance hides a secret and strange way of life. Such is the case for an eel-like fish with ancient origins, a well-known and highly delectable crustacean, and an organism with impressive powers of regeneration that masquerades as an undersea log. The talent among these three marine creatures — the hagfish, lobster, and sea cucumber — is impressive. The hagfish can produce an inordinate amount of slime and tie itself into a knot. The lobster is equipped with supersoaking blasters that it uses to wield a powerful potion; and when under attack, the sea cucumber has defenses that are the envy of science fiction writers. These three organisms are definitely among the oceans' most fascinating and surprising of residents.

THE HAGFISH

To know a hagfish, is to love a hagfish—or maybe not. A good friend of mine in Maine (you know who you are) has developed a new type of phobia; she is convinced that upon entering the Gulf of Maine for a leisurely swim, she will be the target of hagfish. I have tried to convince her that as long as she is not dead or nearly so, they should not be a problem, but she remains unconvinced—hagfish have become her worst nightmare and with good reason.

Hagfishes are blind, jawless, scaleless, and finless fishes with a relatively flexible cartilaginous skeleton somewhat like that of sharks and rays. They resemble eels with a flattened oar-like tail, thick, slippery skin, and one singular nostril above their mouths, around which are several stubby, barbed tentacles. Interestingly, they also have four small hearts.

An adult is typically about half a meter (18 inches) long, though they have been known to reach a scary size of 1.4 meters (4.6 feet). Hagfishes live throughout the world's oceans at the bottom, where it is relatively cool. A few species inhabit shallow waters, but most are found deeper, down to at least 5,000 meters (16,400 feet). It is estimated that there are hundreds of thousands of hagfish residing in the deep waters of the Gulf of Maine.

Though jawless, the hagfish is not without teeth or a means to gain access to tender flesh. It has an extendable tongue equipped with two curved rows of sharp, horny teeth that open and close like a book. Just above that, the hagfish has a fang, which is used to snag prey and keep it from wriggling away. Its toothy tongue and hooked grasp are effective for feeding on soft-bodied creatures, such as worms and other small invertebrates, but not so handy when it comes to prey with tougher skin or scales. Hagfishes have, however, discovered another, easier way to gain access to their victims' tasty, tender insides. They go in through open orifices, such as the mouth, gills, or yes, I am sorry to say, the backdoor. Once inside their prey (already or mostly dead, I swear), hagfishes feast on soft flesh, muscles, organs, and guts. Fishermen know this sly tactic all too well because sometimes upon hauling in their catch all they get is a fish-skin bag full of bones and squirming hagfish.

Along with their gruesome propensity to feed on the dead, hagfishes are well known for their slime, lots of slime (plate 2). If a hagfish, alias slime monster or slime hag, is threatened or injured, it releases mucus from hundreds of glands along its body. In just minutes, one hagfish can fill seven buckets with slime. The glands of the hagfish actually release a thick white fluid containing vesicles of mucus and bundles of thread-like cells. Like balls of string uncoiling, the threads unwrap; they then tangle, combine with the mucus, absorb seawater, and expand into massive amounts of sticky, slimy hagfish goo. Hagfishes use their slime to deter predators and facilitate escape. However, if a hagfish gets caught in its own slime, it can suffocate and endure a most unpleasant fate—death by goo. It has thus evolved a few useful tricks to clear away its own slime. When slime gets up its nose, the hagfish blows it out by sneezing. To free its body of slime, the hagfish wraps its tail around its body and then slides the knot toward its head, scraping itself clear of goo. Its excellent knot-tying skills are also used in feeding to create leverage and improve its flesh-tearing abilities. The hagfish bites onto an irregularity in the skin of its prey and then slides a knot up toward its head, thus enhancing the strength of its pull and ripping power. This process, however, is slow and awkward, so going for the orifices is still a quicker and more efficient means to obtain access to a victim's soft, tasty insides.

Hagfishes spend most of their time at rest hidden within burrows or among rocks at the seafloor. They can also go for long periods of time without feeding. Following the 1989 Loma Prieta earthquake, hagfish at the Moss Landing Marine Lab survived for fourteen weeks without food. They are quick to respond, however, when a meal is at hand and will converge en masse should a bounty of carrion become available at the seafloor, such as a dead whale. Scientists investigating baited traps in the deep sea regularly find them teeming with feeding, writhing hagfish. The hagfishes have excellent olfactory and tactile senses; they readily sniff for and feel out the weak or the dead. Not much to worry about on a leisurely swim, but for me at least, burial at sea is no longer an option. And as for my friend in Maine, she continues to spread the word about the ghoulish hagfish. One of her disciples is a triathlete who competes during the summers in New England when the water is relatively balmy. He regularly dons a wetsuit to ward off not so much the cold, but rather the sneaky hagfish. On a more positive note he says that just knowing that the orifice-seeking creatures are out there makes him swim faster.

Amazingly, there are some organisms in the sea that find hagfishes appetizing—cod and sharks, as well as octopuses, seals, and dolphins make slippery meals of these not-so-lovely fishes. Hagfishes have changed little over the last 330 million years and are thought to be one of the early ancestors of vertebrate animals with a braincase, such as humans. If you thought evolution from primates was hard to swallow, how about having a hagfish in your ancestral lineage?

LOVE POTION #9

They have been called the cockroaches of the sea, were considered junk food by America's early settlers, and are now the ultimate in fine dining. But rarely are lobsters recognized for the power of their pee, their antisocial behavior, or the growing pains they must regularly endure. Over decades of laboratory and field research, scientists have discovered many fascinating, and in some cases rather bizarre, things about lobsters. And a warning if you choose to read on: the lobster on your plate may never look quite the same or quite as delectable.

There are over one hundred species of lobsters found throughout the world's oceans, including the classic large-clawed American lobster, better known as the Maine lobster. There are also spiny, mud, spear, whip, and the shovel-like slipper varieties. Their hues vary, from the typical greenish-brown to tan or red, to almost a bluish color. Due to rare mutations, the well-known Maine lobster can sometimes be found suited in bright blue, white, or an odd half-and-half coloration. The basic body plan of a lobster goes something like this—an external hard shell or carapace, a head that is fused with the upper torso, two stalked, moveable and compound eyes, a tail fin, and ten legs. What chefs and diners usually call the tail is actually the animal's muscle-laden, segmented abdomen. The lobster's firm abs are well toned from use in fast swimming escapes, as anyone who has tried to catch one knows. They use rapid contractions of their abdominal muscles to flap their tails and sprint away backward. At one time, scientists thought that lobsters were mainly scavengers, but now they are believed to be active foragers, and at times, ambush predators. They use their claws, jaws, and legs for crushing, seizing, slicing, and a bit of dicing. On the menu for lobsters are mollusks, such as mussels, clams, and scallops, as well as sea urchins, worms, and crabs. Some lobsters have also been seen to eat fish or filter feed, straining seawater for coarse particulate material. If dead fish are available, they will eat that too, and they sometimes even eat each other. In fact, lobsters have been known to ingest a lot of things, including pieces of plastic, tea bags, wool, and even a rusty nail. In general, however, adult lobsters seem to have a discriminating palate, with a preference for fresh shellfish, crabs, or sea urchins.

Most lobsters, particularly in relatively shallow water, are night owls, nocturnal foragers. Shortly after sunset they leave the protection of their dens to go on the prowl. When they return, often just before sunrise, they may go into the shelter they left from or seek out the closest available place for protection. For the Caribbean spiny lobster, a good hiding hole is best if it also comes with company. It looks for crevices, overhangs, or coral outcroppings that can provide concealment and protection, and that contain other spiny lobsters. More is better when it comes to warding off predators such as sharks or a grouper, as a wall of waving whip-like spines covered with tiny spikes must deter many a hungry invader. A backdoor for escapes is also handy, and many lobster holes have two entrances or exits. The adult Maine lobster, on the other hand, does not seem so fond of its neighbors and will fight fiercely over dominance and the best shelters.

Of course, for these lobsters the best dens are not just good for protection, they also lead to more mates, more sex, and probably more descendants. What determines a winner in the power struggles of the Maine lobster? In this case, size does matter—claw size that is.

Research suggests that the Maine lobster is typically a combatant, promiscuous creature. Undersea battles establish a hierarchy that allows dominant males to get prime real estate and use it to attract the most mates. Posturing and displays resolve some confrontations, while others end up in a brawl, a boxing match, or a brutal fight to the death. Some lobsters choose to avoid opponents all together and will run away from a fight or make a fast retreat with a few flicks of the tail. Lobsters that do decide to engage begin by sizing each other up, whipping their antennae to and fro to feel and sniff out their opponents. They may then push, shove, and lock claw to claw, in an arm-wrestling test of strength. In battle, sometimes it is the lobster that draws a claw first that wins, like a western-style duel, or they may test each other's nerve with a game of "chicken." In the extreme, claws or other appendages may be torn or ripped from their bodies. Luckily, lobsters can regenerate most of their appendages. If an eye is lost, however, they cannot grow it back, and strangely enough another appendage may grow in its place, such as an oddly located walking leg. And if the need should arise, a lobster can jettison or slice off its own limb, a clever escape tactic, especially if you can grow back the lost appendage. In large tanks, some victorious lobsters have been observed to show mercy on the defeated, while others are not so kind and may mutilate or hack the loser to death. Research has also revealed that in fights, lobsters get really pissed off—literally.

Whether it is as a precursor to battle or in a bit of foreplay, when Maine lobsters meet, pee matters. They are well-equipped and stocked to make good use of their urine. Each lobster has a pair of muscular nozzles located just below its antennae—a twin set of built-in pee-blasters, which are connected to an ample supply of urine that is stored in two bladders also located within its head. To further its pee-shooting range, a lobster can generate water currents with its gills and mouthparts, enabling it to reach a target, such as an opponent's face, some seven body-lengths or about 1.5 meters (5 feet) away. Lobsters actively sniff for undersea "odors" or chemicals by flicking their smaller pair of antennae, or antennules, back and forth. In laboratory studies, the lobster to pee first and with the "sweetest" smelling urine, along with the largest claw, is the most likely to win in battle. Underlying the effects of the lobster's urine are hormones that seem to control aggression and additional chemicals or pheromones that act as this leggy crustacean's version of "Love Potion #9."

When a female Maine lobster approaches a shelter, hot for some action, she not only sniffs for a male's pee, she lets loose a stream of her own. Her urine can render a once brutish male docile and even touchy, feely. Instead of smacking the female over the head with his crushing claw, the seduced male waves his antennae gently over her body as she enters his den. On occasion, a female's love potion may not be fully effective and she may be rejected, especially if she is unprepared to come out of her shell—and not in a metaphorical sense. Before mating, a female lobster molts, whereby she becomes soft, vulnerable, and her relevant private parts are accessible to the male. By doing so, she also conveniently provides her mate with a nutritious postcoital snack, her molted shell. After mating, a female lobster may spend a few days recovering from her molt within the male's shelter. She then simply walks away and a new female lobster will come to call. Dominant males are repeatedly seduced into a continuous series of short-term affairs, while the females seem to choose when and with whom they will mate. The subordinate males, those that do not win battles or get the best lairs, will sometimes get a few of the dominant males' leftovers, but without a large, attractive condo to share, they remain mainly frustrated bachelors on the make.

Female lobsters can store the males' sperm for up to about three years, using it to fertilize several batches of eggs. They may carry tens of thousands of eggs glued under their abdomens for some ten months before hatching occurs. In their larval stages, the young lobsters join the ranks of the plankton for days, weeks or possibly months depending on the species and surrounding conditions. Each baby lobster will go through several developmental stages before growing into its more familiar form and taking to a life at the seafloor. Juvenile lobsters tend to live in shallow, protected coastal habitats until they are large enough to safely roam at greater depths.

For lobsters, molting is an important part of mating, a life-long necessity, and conceptually at least, a painful process. Like other crustaceans, as a lobster grows it must molt to replace its rigid carapace with one that is larger and able to hold a bigger body, sort of like turning in a small compact car for a minivan. Mature lobsters may molt several times a year; juveniles must do it more often because they grow faster than the adults. But they don't simply leap out of one shell and grow another; it is a lengthy, fascinating process, and lobsters spend a good part of their lives undergoing the changes involved. Molting begins with some serious dieting, as a lobster must shed some of its mass. Simultaneously, a new paper-thin exo-skeleton starts to form under its shell and its blood is moved from its outer appendages, like the claws, spines, or legs, into its body. Then it is time for a drink, a really big drink. A lobster guzzles water so that its body swells and its old carapace is pushed apart. Essentially, some serious bloating causes the lobster to unzip, unhinge, and literally burst at its seams. Lying on its side, slicked up with some lubricating slime, a lobster then must pull its body, including the antennae, legs, spines, claws, and mouthparts, out from the remains of its old shell. For Maine lobsters, particularly those well endowed in the claw area, the process must be especially difficult and possibly painful. They must pull their large, bloated claws through the slender jointed wrists of the old carapace. Think of trying to squeeze swollen hands through a pair of handcuffs—and they have to do it every time they molt. Once its appendages are through and the last bit of its shell has been shaken off, the newly emerged lobster or "shedder" is a floppy, jelly-like creature trying to stand up on wobbly legs with a shell the consistency of thin, wrinkly plastic wrap. It then goes again for the bloat; drinking water to inflate its size even further so that it has room to grow within its new carapace once it has hardened. A lobster typically devours some of its old shell for a megadose of minerals and nutrients.

Shedding can take just several minutes or last for up to half an hour. It is a dangerous time for the lobsters, as they are immobile and defenseless. They may go into seclusion for several days, emerging only after their new shells have begun to harden. The first body parts to stiffen are those most critical to foraging, such as the tips of the walking legs and mouthparts. It can take several months for the lobsters' carapace to harden completely. Maine lobsters molt principally in the relatively warm summer months.

The Caribbean spiny lobster may not have the brutal mêlées or social rankings of the Maine lobster, but they exhibit at least two very curious and unique behaviors. Just after the first autumn storm, in locations such as the Bahamas, Florida, Cuba, the Gulf of Mexico, and Central America, spiny lobsters begin a two-to-three-week trek into deeper, offshore waters. Many marine organisms make lengthy migrations, some much longer than that of the spiny lobster, but few others do it with such style. During their fall trek, thousands of lobsters will traverse the open bottom, marching in an amazing single-file formation known as a queue. They line up head to tail, each lobster closely following the one in front, guided by the touch of its antennae (plate 3). The movement of the lobsters seems to entice others to leave their shelters and join the crustacean train. Scientists think that a queue is formed to reduce drag, like a professional bicyclist drafting behind the racer ahead. It may also help to prevent predation or aid in orientation while marching. The spiny lobsters are thought to venture into deeper water to avoid the relatively cold temperatures brought on by storms in the fall and winter months. Other lobsters are known to migrate seasonally, between shelters and habitats, but the spiny lobster may be the only one that creates a single-file offshore express. Experiments suggest that lobsters use the Earth's magnetic field as a guide to navigate the open ocean and that chemical signals may lead them to specific home ranges or locations.

The spiny lobster's acute sense of smell also appears to provide it with an exceptional medical diagnostic capability; one that doctors can only dream of. Mark Butler, a professor at Old Dominion University, and Donald Behringer, a research scientist at the University of Florida, discovered that juvenile spiny lobsters will actively shun diseased neighbors. This normally social lobster will avoid dens that harbor lobsters that are infected with a lethal, pathogenic disease, essentially placing them in quarantine. Even more startling is that the juvenile spiny lobsters seem to be able to detect or "smell" the disease before it becomes infectious. Butler suggests that their behavioral change is an adaptation to thwart the spread of a lethal disease, and that it may be the only known example of this sort of "shunning" in the animal kingdom.

Even with their crushing claws, spiny swords, shield-like carapace, and of course, Super Soaker pee blasters, lobsters are not invulnerable to predators. A wide range of creatures find lobsters fine dining, including fishes, sharks, sea turtles, octopuses, and of course, the most feared of all—humans. For those of you who like to eat the disgusting gooey green stuff inside a lobster's body, the tomalley, it is the liver and pancreas combined, which acts as a filter and can accumulate pollutants or toxins over time. It is probably best to forgo this rather questionable delicacy.

SHAPE-SHIFTERS

The humble sea cucumber sits like a log on the ocean floor and may resemble its namesake from the garden or grocery store, but this creature is far from a simple lackadaisical lump; it is endowed with some very special abilities. Sea cucumbers are tube-shaped and come in various colors, such as tan, green, or black, with bumpy, leathery skin, growing to a typical length of about 15 to 25 centimeters (6 to 10 inches). That's the garden-variety sea cucumber. In fact, there are more than a thousand described species, and many of them are more of a nature-gone-wild version of their fruit look-alike. That's right, cucumbers are in fact a fruit, not a vegetable—one never knows what interesting fact will turn up when researching a book! Sea cucumbers may be dressed in psychedelic hues of electric blue, vivid purple, or shocking red, and sport spikes or frills or appear almost furry. The giant red or California sea cucumber is the largest of its kind, reaching a monstrous length of nearly a meter (30 inches). My favorite is the chocolate chip sea cucumber, covered with tan skin folds resembling that of a shar-pei puppy and speckled with black-brown spots. Sea cucumbers are found worldwide, ranging from the deep sea to shallow shores, and can live in mud, in sand, on rocks, or in coral reefs. They attach to the bottom or move about sluggishly on five rows of small tube feet, which are extended or retracted using an internal hydraulic system of seawater.

For sea cucumbers life is all about finding yummy particles of organic matter. Some sea cucumbers filter seawater to acquire these tasty bits; others use mucus-covered tentacles that they raise up or sweep over the seabed. They bring their tentacles into their mouths, lick off the entrapped particles, and then re-release their sticky collectors to gather more food. There are also some slurpers; these sea cucumbers crawl over the seafloor slurping in sand or mud to sift out the organic matter. It is when threatened that the sea cucumber reveals its truly odd nature.

Sea cucumbers are real-life shape-shifters; if predators near they can morph their skin from hard and lumpy into something a bit less appetizing, akin to a gelatinous slime. When danger looms they literally turn to mush! Sea cucumbers have another very effective and rather disgusting means to deter predators: they eviscerate, readily expelling their insides as a decoy or trap. Some sea cucumbers eject a sticky spaghetti of white tubules, while others release their actual internal organs. A predator such as a fish or sea star may become entangled in the slimy mass or be distracted long enough for the sea cucumber to slowly crawl away. Imagine the shock on a mugger's face if on demand you could let loose of your insides. Amazingly, not only does the sea cucumber survive, but in just three to five weeks it also regenerates its internal organs. Unless you like a handful of guts, it is unwise to pick up or harass a sea cucumber. Some sea cucumbers also exude a toxin, which can leave aquarists baffled when they add this seemingly peaceful creature to their tank and a mass mortality ensues.

Relatively recent research has also discovered that in the deep ocean there are strange, translucent sea cucumbers that can swim—well, sort of swim; actually it looks more like flying. They have specialized wing-like flaps at the front and/or back (often difficult to distinguish, by the way) and can lift off, swim a ways and then land back on the seafloor. This adaptation is thought to allow them to more efficiently find food in the deep sea, where dining options are often quite limited and may come in periodic windfalls from above.

I queried a group of graduate students studying marine biology at the University of Miami's Rosenstiel School of Marine and Atmospheric Science about what creature they thought was the oceans' most unusual. One young woman responded immediately: her nomination was the pearlfish and its unique, rather bizarre relationship with sea cucumbers.

The pearlfish's association with the sea cucumber hinges on the fact that the sea cucumber breathes through its butt. Water flows in through the "back entrance" and washes over the animal's respiratory organs, wherein oxygen is consumed and carbon dioxide is released. The water then flows back out the way it came in. The tiny pearlfishes have handily evolved the ability to detect the chemical signature of the sea cucumbers' respiratory outflow. This becomes useful for a pearlfish after a night of foraging, when seeking a protective daytime shelter. Like a heat-, I mean butt-, seeking missile, the pearlfish uses its detecting skills to find a sea cucumber and enter via the backdoor. Its hiding spot within the sea cucumber comes with an added benefit; it is already well stocked with provisions. The pearlfish nibbles on its host's respiratory or reproductive organs. Understandably, some sea cucumbers are not willing hosts and will eject their respiratory or digestive organs to deter the sneaky little fishes. One species of sea cucumber goes even further, as it has evolved a deterrent: a backside lined with teeth to fend off the crafty pearlfish.

WHY THEY MATTER

Sea cucumbers, hagfishes, and lobsters all play important roles in the ocean ecosystem and have both obvious as well as more subtle connections to our everyday lives. Within the sea's web of life, hagfishes and lobsters play duel roles, as both scavengers and predators. As predators, these animals keep prey populations in check and remove the weak or sick from the gene pool. As consumers, they also transfer energy in the form of carbon (organic material) up through the ocean ecosystem. Lobsters and hagfishes also provide sustenance for the organisms that eat them, such as fishes, sharks, or marine mammals. Sea cucumbers provide nourishment for other organisms as well, including sea turtles, sea stars, some crustaceans, and many fishes.

As scavengers, the lobster, hagfish, and sea cucumber are part of the oceans' cleanup crew. The hagfishes are probably the best of the bunch, providing a rapid and effective means of cleaning up the dead and rotting of the sea, anything from a 100-ton whale to the discards from an industrial fishing ship. In fact, the practice of dumping fishery discards, the unwanted or too small, is believed to have fueled an increase in some hagfish populations. Sea cucumbers tidy up both the water and the sediments. Where filtering sea cucumbers are abundant, they play an especially important role in promoting water clarity and quality. Slurping sea cucumbers strain organic matter from the sediments. Much like earthworms in a garden, sea cucumbers and other creatures that burrow or feed within the sediments of the seafloor also help to keep the bottom aerated and well mixed.

Food

For most of us, lobster certainly sounds more appetizing than a nice plate of hagfish or sliced-up sea cucumber. In Asia, however, people eat both hagfishes and sea cucumbers. Imagine the eel-like hagfish skewered in s-shaped folds and then roasted—slime hag on a stick. Or how about a nice dinner of braised sea cucumber or a side order stir-fried in black pepper sauce? The fermented viscera of sea cucumbers, aka pickled gonads and intestines, is considered a delicacy in Japan. People tell me that sea cucumber is rather plain tasting, and that much like tofu it sucks up the flavor of whatever it is cooked in. One culinary advisor even suggests that no gourmet should go without experiencing the succulent jelly-like texture of a sea cucumber. I have to question such advice on a whole number of levels.

When it comes to more popular seafood, the lobster is a tasty icon. The variety of ways in which it is served seems endless—the classic steamed lobster, the lobster roll, baked, stuffed, on the grill, covered in a rich cream sauce, stir-fried, poached, and now, as a means to make mac and cheese an epicurean delight. Lobsters are sold throughout the world; they can even be shipped directly to your home. It is a must-have for luxury hotels, restaurants, and cruise lines. Lobster meat is also low in fat and cholesterol—it's the butter for dipping that will go to your waistline and block your arteries. But lobster has become more than just a simple food; eating this ocean animal has become synonymous with being able to afford the best and an important part of a cultural, regional culinary experience. No matter how it is cooked or served, lobster is about much more than just food.

The state of Maine accounts for most of the lobster caught in the United States, with landings in 2008 valued at more than three hundred million dollars. A drive along Maine's coast is all that is needed to see the importance of lobster to the region. Images of the clawed crustacean are everywhere, on signs for roadside seafood shacks and fancy restaurants or hotels, decorating windows, doors, and mailboxes. In this area of the world, lobsters are like hot dogs to baseball or chips to salsa; they are the underpinning of tradition, culture, and tourism. People from around the world travel to Maine to partake of lobster pulled fresh from the sea. Thousands of people are employed directly and indirectly by the industry in fishing, processing, marinas, shipping, hotels, souvenir shops, and in restaurants. And here, lobstering is serious business: battles for fishing territory can get ugly, and illegal harvesting is a grave offense. For the fishermen involved it is a way of life that has passed from one generation to the next and which they fiercely protect. I was fortunate enough in 2009 to spend a day out on a lobster boat off the coast of Maine with Tommy, a seventy-eight-year-old fisherman. As he pulled his traps, I was the bander of the lobsters' claws and helped with rebaiting. It was a rainy, rough, windswept day on the seas, which in no way deterred Tommy or his love for the job. I asked what kept him going day in and day out, hauling hundreds of traps each day for so many years. He said it was simple for him and many others like him; it was a love for the sea and a curiosity to see what each trap would reveal. Midway through the day I understood his calling, peering eagerly over the boat's rail as each trap was hauled up. Would there be a fish, crabs, big lobsters, small lobsters, or females with eggs that needed to be thrown back to preserve the population? It was an endlessly fascinating and productive day. Hopefully, lobsters will continue to provide for this ocean-going, ocean-loving way of life, one that so defines and supports coastal Maine.

Human Health

Here's a surprise—hagfishes may prove beneficial to human health. That's right, the hagfish. Scientists are very interested in their primitive immune systems, which have long protected them against infection. It is especially intriguing given the hagfishes' propensity to feed on the dead, which are usually laden with bacteria and other microbes. Researchers have already identified three potent, broad-spectrum antimicrobial compounds in the Atlantic hagfish that might help to explain their ability to ward off microbial diseases. Hagfishes have also been found to get liver cancers and may provide a means to monitor carcinogenic pollutants in the marine environment.

Sea cucumbers are proving useful in biomedical research as well. Scientific studies have shown that a protein found in sea cucumbers may be an effective tool to inhibit the development of the malaria parasite. Other research suggests the use of sea cucumber extracts to fight colon or pancreatic cancer. Though scientifically unproven, compounds from sea cucumbers are also promoted in dietary supplements as a treatment for arthritis. One only has to wonder about how they regenerate their internal organs to ponder what other medical applications sea cucumbers may one day provide.

Rather than eating lobsters, some scientists are using them to learn more about biology and physiology. Researchers are studying their extraordinary ability to sniff out food, rivals, and the opposite sex to improve our understanding of the sense of smell. Investigators are also looking at the optics of the lobsters' eyes and studying their nervous system and how it controls locomotion and other bodily functions.

Other Uses

In Korea, the deslimed skin of hagfishes is used to make "eel skin" products, such as handbags, shoes, wallets, and briefcases. The hagfish's slime may in and of itself prove of value in biotechnology. Researchers are investigating hagfish "slime threads" as analogs to spider silk. Douglas Fudge of the University of Guelph explains, "We're targeting high-performance applications that could replace polymers like Kevlar, but ultimately, we want to replace materials that are taken for granted every day, like polyester, polypropylene, and nylon, with those made from renewable resources." Just think—one day you could be wearing or using biodegradable material made of synthetic hagfish slime. In New Zealand, hagfish slime is also reportedly used by the Maori as a cleaning agent.

The secret lives of the hagfish, lobster, and sea cucumber provide excellent examples of the weird and wild things going on within the sea. These three organisms also illustrate the importance of even the most bizarre creatures in keeping the oceans functioning and healthy. And what creature better than the lobster to demonstrate our long-standing and strong ties with the sea and marine life, though all three animals are showing great promise in biomedical research. The ocean as well as society benefit when the sea is replete with an abundance of slime-touting hagfishes, shape-shifting sea cucumbers, and pee-wielding, seductive lobsters.

3 Let's Talk Snails

The highly prized gastropods we call escargot owe their renown to delectable baths of garlic butter, but in the salty brine of the sea, there are equally illustrious snails that are celebrated for far more than just gastronomic pleasure. One small marine snail with wing-like feet uses a bubble of mucus as a floating flytrap to capture its food and has been called the potato chip of the sea. Another of the oceans' gastropods is exceptionally large, famous for its lustrous pink shell, and has some unusual, shall we say lengthy, reproductive traits. One of the most beautiful but deadly creatures in the oceans is also a snail. And finally, there is one shell-less, naked snail that possesses chemical weapons and has cannibalistic tendencies that make even its sexual liaisons literally life threatening. It is also one of most colorful animals on the planet. The oceans' resident snails are rich in diversity, full of surprises, and live in a world of slime, sexual promiscuity, and warring factions.

FLYING SNAILS

Pteropods are small, weakly swimming snails that are found in all of the world's oceans, from pole to pole, typically in the upper 200 meters (656 feet), though some can live deeper. The greatest variety of pteropods is found in warm water, but it is in the cold that they flourish in abundance. One species even lives full-time under the ice. They are a far cry from the stereotypical snail. Rather than dull creatures relegated to an existence of sluggish crawling about the bottom, the pteropods are beautiful, active animals that spend their entire lives in the water, floating, drifting, swimming, eating, and of course, mating. For fans of the undersea sci-fi classic

The Abyss, a pteropod seems a likely inspiration for the ethereal alien creatures that were the saviors at the film's end.

The muscular foot that most snails use for locomotion has been modified in the pteropods into delicate fins for swimming. Some species have paired fins, while others have a large, rounded or heart-shaped plate, like a giant flipper. Using a flapping or sculling motion, a pteropod can produce short bursts of speed for evasive action or engage in long, slow swims upward at night. They also swim to prevent sinking and to maintain an advantageous position during mating. As relatively weak paddlers, though, the pteropods are considered zooplankton. Some pteropods have shells, while others go without, and they range in size from about 1 millimeter (less than 0.25 inch) up to about 8.5 centimeters (3.4 inches).

To envision a typical shelled pteropod, imagine a snail with a small coiled or cone-shaped shell, upside down with little, fluttering, transparent wings extending from the top. Pteropod shells are thin to reduce sinking, are composed of aragonite (a form of calcium carbonate), and come in a variety of shapes, including whorled, straight, pyramidal, and rounded. Some pteropods have an internal pseudoshell that can be jettisoned if attacked. One bizarre species has an elephant-like trunk and two large side wings, each trailing a long tentacle. For the most part, pteropods are transparent, though some have pigmented spots. One advantage for these pteropods is that when predators near, they can withdraw inside their shells for protection.

For the shelled pteropods, feeding is a very sticky story. They use an ingenious method to capture plankton, and it comes with an added undersea benefit—extra lift. These wily gastropods create large, transparent webs or bubbles of mucus that extend above and beyond their bodies like parachutes. A pteropod may hang motionless or sink very slowly for hours, as small diatoms, foraminifera, or dinoflagellates become trapped within its expanded bubble of goo. Once replete with yummy morsels, the pteropod simply slurps up its slimy net and chooses which particles to dine on. If disturbed, a shelled pteropod may ditch its sticky parachute, leaving it as an easy feast for another hungry denizen of the sea.

Pteropods without shells are sometimes called sea angels due to their appearance. Each has a small head and short neck atop a narrow, bag-like body with little, pointed wings extending from its sides (plate 4). At the front of its head is a pair of cute, knobby tentacles. These animals are agile and relatively speedy swimmers; they can take evasive action if predators near and are good hunters in their own right. The naked pteropods are also picky eaters and prefer to feast mainly on their cousins—the shelled pteropods. They are well equipped to capture and consume their relatives with sucker-bearing arms, hooks, or spiny jaws. They can also be very determined consumers, and if needed, will wrestle for hours to pry a victim from its shell. Once extracted, the soon-to-be-eaten morsel, aka its cousin, is swallowed whole and the shell tossed aside. When hungry, the shell-less pteropods will also steal food from their neighbors or, if necessary, eat other types of zooplankton.

As for their sex life, it is active and free of gender issues. Many pteropods start as males, mate male-to-male, and then turn into females. During mating, male pteropods exchange capsules of sperm, which they store for later use as females. The deed itself is a tricky midwater endeavor, given their propensity to sink. Pteropods sometimes float, often swim, and may hold on to each other as they mate. And they have different position preferences—some like to do it back-to-back, while others prefer a more traditional face-to-face arrangement. After mating, a pteropod's male organs are reabsorbed into its body and female parts develop. Then, with sperm already in hand (or body as the case may be) and the right equipment now in place, most pteropods lay thousands of fertilized eggs within long, floating ribbons. A few pteropods brood their eggs, releasing a smaller number of more mature offspring. In several of the shell-less pteropods, their private parts are adorned with spines and they have a suckered, tentacle-like appendage to aid in copulation. Some of these winged creatures may function as males and females at the same time, and pairs have been observed in an endurance-testing embrace lasting for up to four hours. It is an energy-sapping process as they slowly beat their fins to stay afloat. Imagine an intimate encounter that lasts for hours and all the while, you and your partner must tread water. To stay fueled up, pteropods may continue to feed while mating, like grabbing a sandwich on the side during the action.

Pteropods are relatively easy prey for larger creatures, and their abundance has led the shelled variety to be likened to the potato chips of the sea, a bit of a crunch in a fast-food meal. The swimming snails are an energy-rich snack, complete with protein and fats. They are scooped up by whales, captured by jellyfish, plucked out by seabirds, and crunched on by seals, fishes, squids, and the ever-voracious arrow worm. In the Antarctic, there is a specific species of pteropod that may also fall victim to kidnappers. An amphipod there is known to abduct these small snails and carry them clasped in a set of pincers. The species of pteropod taken hostage is believed to have a chemical defense against some Antarctic fish; thus the amphipod shrewdly uses it as a biological weapon against potential predators.

SNAIL ROYALTY

The popular queen conch (pronounced "konk") is an exceptionally large snail with a beautiful pink luster to its shell, and a male with an unusually lengthy reproductive "sword" or verge. The mature queen conch has one of the most well-recognized live-aboard homes in the world; a large whorled shell with a flared lip whose dull, tan outside belies a glossy rose sheen within. It is found in warm waters from Bermuda to South Florida, the Bahamas, throughout the Caribbean, and off the northern and southern shores of Central and South America. Although queen conch like to hang out in areas of sand or seagrass, they may also be found on hard bottoms or among coral rubble. Beneath its mega-shell, the conch has two yellow, stalked eyes along with a long, retractable snout or proboscis that it uses to graze mainly on algae and seagrass. The conch also has an interesting aid to digestion—the crystalline style, which is a transparent, gelatinous rod that rotates while slowly being pushed into the animal's stomach. The rotating rod acts like a cocktail stirrer and releases digestive enzymes as its outer layers are dissolved. Native lore suggests that the crystalline style is an aphrodisiac when consumed, although there seems to be no scientific evidence to back this up. Personal experience suggests a lack of taste as well.

The queen conch's favored mode of transport seems somewhat unusual for a snail; rather than slither or crawl, surprisingly, it often hops. Extending its muscular black-speckled foot, a conch fixes itself to the bottom, and then pulls or throws its heavy shell forward (or backward) via a vigorous contraction, jumping from one spot to the next. At the end of its foot, the conch has a sickle-shaped horny covering, which can be stuck in the sand and used to improve its long-jumping ability. The conch also uses its extendable, strong foot when overturned, to right itself.

Juvenile queen conch have well-developed, thin spines on the spire of their shells and begin growing a broad, thin lip. After sexual maturity, at about three to three and a half years of age, the conch's shell no longer grows in length; it will, however, thicken with time, and its spire's spines will get worn down. Sponges may bore into the outer layers of the shell, and it usually acquires a furry overcoat of algae. A mature conch only adds to or does repairs on the inside of its shell; consequently as the animal ages it must shrink to fit within its downsized home. Old conch, called "samba," look so different from juvenile or young animals that at one time they were thought to be a different species; we now know they are just the senior citizens of the group. Conch can live up to twenty or thirty years if undisturbed and in a favorable environment.

Now back to the infamous verge (plate 5). Conch biologist Allan Stoner, a senior scientist at the National Marine Fisheries Service, says, "There is a distinct advantage to studying the reproductive biology of an animal that is big, slow, mates for hours on end, and includes a male with a penis nearly half of its total body length." In queen conch, reproduction may occur year-round or be the inspiration for a bit of travel and a party. Some migrate to a specific area each year for what is essentially a big snail orgy in the sand. And the more the better; reproduction in queen conch is believed to be density-dependent, so that if not enough individuals come to play, mating and the production of offspring are unlikely. Research has also revealed that male conch are not only well endowed, they are also lustful, rather undiscerning, and, from a female's perspective, a bit inconsiderate. Tracking the females probably by pheromone-laced trails in the sand, the males approach from the rear and set their shells behind the objects of their desire. Now consider the size of their shells, potentially reaching over 20 centimeters (8 inches) in length. The male conch must be able to extend his delivery tool out of his shell and under that of the female—hence the very long verge. It is a thin, black, spade-tipped, and agile projection that the male conch uses with alacrity. Once done with one female, he moves on to the next. The only problem is (the squeamish may want to skip this section) that when extended outside his shell for the purpose of copulation, the verge is unprotected, and hungry crabs and eels are only too happy to take advantage of the male conch's vulnerabilities. Luckily though, when he loses one, the male conch just grows another—the conch is able to regenerate its penis!

An enthusiastic male may continue to ply a female conch with its verge even while she is laying eggs. Hundreds of thousands of eggs may be laid at one time in a bundle of gelatinous strings on the sand. Some five days after spawning, the eggs hatch and larvae conch are released into the water, where they remain as plankton for about two to four weeks before settling to the seafloor.

Anyone who has ever tried to clean a queen conch knows about one of their other distinctive traits—they are coated in a sticky, gooey slime that

PLATE 1. A floating solar plant and animal all in one. A foraminifera with a calcium carbonate shell (0.5 mm across), spines, and arms of goo with its algae cells deployed to capture sunlight. Photo by Dave Caron, University of Southern California.

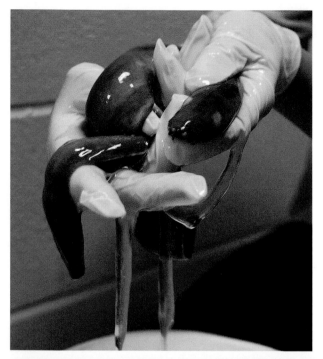

PLATE 2. Researcher holding the mega-slime-producing hagfish. Photo courtesy of Tim Winegard.

PLATE 3. The crustacean train, the single-file march of the spiny lobster. Photo © Doug Perrine/SeaPics.com.

PLATE 4. A shell-less pteropod or winged snail. Photo by Russ Hopcroft, University of Alaska/NOAA/Census of Marine Life.

PLATE 5. Conch sex and the male's lengthy verge revealed. Photo by Jerry Corsaut.

A

B

PLATE 6. The outrageously flamboyant and diverse sea slugs, or nudibranchs. (A) Mating pair of clown nudibranchs. Photo © Diane Armstrong/SeaPics.com. (B) Mediterranean nudibranch. Photo © Franco Banfi/SeaPics.com. (C) The "Fried Egg" nudibranch laying an egg mass ala slime. Photo © Mark Strickland/SeaPics.com.

C

PLATE 7. Close-up of coral polyps extended at night to feed; one has captured a small worm. Photo ©
Doug Perrine/SeaPics.com.

PLATE 8.
A moray eel
opens wide to
breathe. Photo
by Karen Doody.

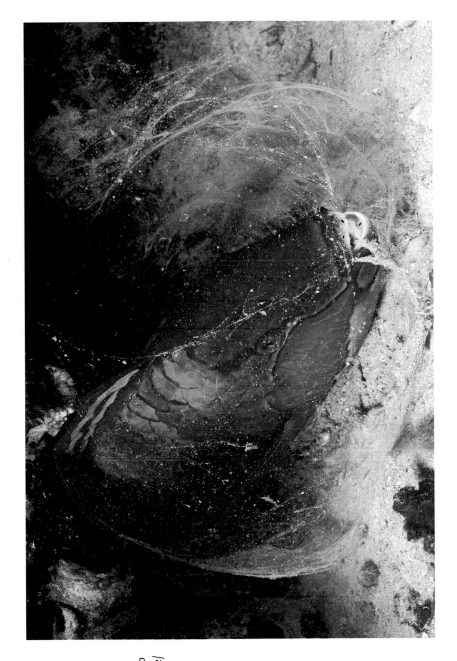

PLATE 9.
A parrotfish
slumbers in a
cocoon of pro-
tective slime at
night in the
Celebes Sea,
Malaysia. Photo
© Doug Perrine/
SeaPics.com.

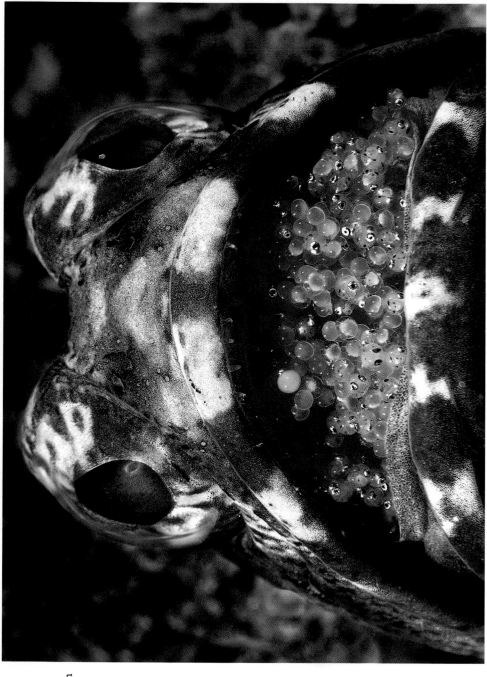

PLATE 10.
A male jawfish
with a mouth-
ful of eggs.
Photo by Steven
Kovacs.

rivals the mucus of the hagfish, if not in quantity, certainly in its adhesive, difficult-to-get-off, properties. The conch is thought to use its slippery mucus to slicken its path and as a means to ward off potential predators.

Even with the armor of a thick shell, a muscular foot with a pointed, hard covering, and an abundance of slime, the conch is still fair game for some of the oceans' better-equipped hunters. The queen conch's cousins, the tulip shell and horse conch, are ones they must be wary of. Other predators that can either pry the animal out of or crush its shell are the octopuses, spiny lobster, spotted eagle ray, sharks, sea turtles, and several types of fishes. Juvenile conch are particularly vulnerable and spend much of their time buried in the sand, feeding primarily at night. Humans are, of course, the most deadly of the predators that feast on the queen conch.

KILLER SNAILS

A suave secret agent walks into a plush bathroom at an elegant hotel where he is recovering from his most recent mission, having just thwarted the latest attempt at world domination. Walking toward the shower and the warm water that will soothe his bruised and broken body, he notices a beautifully patterned seashell on the floor. He is wary and looks around for evidence of foul play, searching the room for uninvited guests, such as the poisonous snakes and deadly spiders that have been previously used in attempts on his life. He finds nothing else out of place and sees a few, similar, presumably decorative shells in disarray on the sink. Breathing a sigh of relief, he bends down and nonchalantly picks up the shell to place it back on the sink from which it must have fallen. Big mistake! A harpoon like tooth juts out from the seemingly innocuous shell, piercing his palm. His hand is wracked with pain as it begins to swell and turn red. Lunging into the bedroom, he grabs his jacket from the bed, desperately seeking the antivenom injector pen in the pocket. He is quickly weakening, and as the first tingling of paralysis hits, breathing becomes difficult. The once invincible secret agent has finally met his match—the killer cone snail.

There are over seven hundred species of venomous cone snails in the ocean. They inhabit temperate, subtropical, and tropical regions and are especially prevalent in the Indo-Pacific. Cone snails range in size from a few to 20 centimeters (8 inches) in length. They may be found in coral or rocky reefs, muddy mangroves, or intertidal lagoons. They are all hunters and are most active at night. As ambush predators, cone snails lie in wait, buried in the sand or mud, under rocks, or beneath coral rubble. A tubular siphon allows them to breath even when buried and is used to chemically detect their prey. Some cone snails use their snout or proboscis as a lure for the unwary. The cone snails that preferentially strike and immobilize the fastest prey, fish, are the most lethal. Scientists have described the venom of those that hunt fish as a pharmacological brew that is like a mix of the curare-laced poison Amazon Indians use on arrows and the nerve toxin of the fugu pufferfish with a lethal dose of botulism thrown in. While some of the killer cone snails specifically hunt and dine on fish, others prefer worms or mollusks.

The cone snail is extremely well equipped to capture, kill, and consume prey that are faster, bigger, and maybe even smarter. They have a keen sense of smell with which to ferret out potential prey and an armament of venomized teeth. A sac at the base of the cone snail's snout contains two bundles of spear-like teeth. One bundle is an armament reserve in which teeth are generated and held as backup. The other bundle or part of the sac holds several fully developed and hollow teeth ready for use. The shape and size of the teeth varies with species, with the largest reaching up to 2 centimeters (almost an inch) long. Some teeth are barbed, while others have serrated edges. In the attack mode, the cone snail extends its snout toward a victim and moves a tooth forward, loading it with venom (scientists are not sure how this happens). The hunter then spears its victim and reels it in by a tether that is attached to its tooth. Paralyzing venom may be injected right away or only after the prey has been engulfed within the cone snail's expandable snout. And if one venom-laden harpoon isn't enough, the cone snail reloads and fires another. The cone snails that preferentially hunt shellfish use their venom to induce paralysis, which makes it much easier to pry their prey out of its shell. Researchers have discovered that cone snails have a tremendous variety of venoms and amazingly, individual organisms can even change the composition of their toxins between strikes. Their at-the-ready harpoon-like teeth and array of varying and potent venoms more than make up for this snails' lack of speed and agility.

Some thirty human fatalities have reportedly occurred due to encounters with cone snails, partly because their venom acts so rapidly and because to date there is no specific anti-venom available. The fish-killing and largest cone snails are the most dangerous. It is difficult to distinguish, however, the varieties that cause only slight pain from the one-nick-and-you-are-dead type. In other words, don't pick up a cone snail unless you really, really know what you are doing. As their name suggests, all have a cone-shaped shell. Experts differentiate species based on the color, shape, and texture of the shell along with the structure of its whorl. But these characteristics vary not only between species, but also within a species. Further complicating the picture is that some cone snails are covered by a thin layer of tissue that prevents overgrowth and provides camouflage, but also makes discerning the color, pattern of the shell, and thus the species, aka deadliness, more difficult.

Due to their natural beauty and tremendous diversity, the shells of the cone snail are highly prized by collectors. Shells may be white with black or brown markings, or blue, dark violet, orange, or red. They can also be glossy or dull, ridged or smooth, and thick or thin, as well as decorated with zigzags, spirals, dashes, or dots, or vertically striped like a zebra's hide. They may resemble the patchwork appearance of snakeskin, the artistic designs woven into African baskets, or the tie-dyed batiks of the Caribbean. "Beautiful but deadly" is a most fitting description for these, the oceans' lovely, yet lethal gastropods.

Little is known about reproduction in cone snails, but it is safe to say that it must be done with care. As for what creatures dare to prey on the killer cone snail, for one, other cone snails, although they appear not to be immune to their own toxins. Crabs and some fishes have also been known to successfully capture and consume them—without dying.

NAKED SNAILS

In the oceans, the sea slugs have it all—diversity, beauty, sex, slime, and ingenious self-defense. These are not the humdrum slugs of your garden. Their most obvious and striking characteristic is their fantastic coloration, but there is much more to these naked, soft-bodied creatures than just their flamboyant attire.

There are thousands of species of sea slugs living throughout the world's oceans, with the vast majority residing in the tropics. Most are naked, snails without shells, though some have a reduced or internal shell. They may be small, just a few centimeters in length, or as in California's gigantic sea hare, *Aplysia vaccaria*, grow up to 1 meter (3.2 feet) long and weigh a hefty 30 pounds!

To envision the basic body plan of a sea slug, think of a snail sans shell. It has a soft, slightly elongated body with a well-developed flat foot underneath for crawling. The head and tail may be either distinct or hard to discern from the rest of the animal. The eyes of most sea slugs are little more than vague or indistinct pigment spots, which may detect variations in light. A distinctive characteristic and one that helps tell the front from the back of a sea slug is a pair of short sensory tentacles on the head, the rhinophores. These are the sea slug's nose, helping it to detect friend, foe, and food. In one group of sea slugs, the sea hares, the sensory tentacles look like ears, specifically rabbit ears. At the front of the head near the sea slug's mouth is another pair of small, stubby appendages; these are more sensitive to touch and taste. The gills of the sea slug are also a distinctive feature and may be located within the body cavity or along an animal's sides. Sea slugs that have an odd "naked gill" structure sitting peculiarly atop their backs are called nudibranchs.

To the basic sea slug body plan, add seemingly infinite variations in shape, color, texture, ornamentation, and gill and rhinophore structure (plate 6). It is as if Mother Nature asked a children's art class to design the sea slug and gave them instructions to use all the colors of the rainbow, to attach any sort of paper cutout for decoration, and most especially, to use their imaginations with abandon.

Some sea slugs are classically elongate and slender, while others are more oddly shaped, resembling a small oval pancake or a tube with a bulging, round middle. The most striking characteristic of the sea slugs is their color; just imagine the most vivid hues and wild patterns possible. The polka-dotted and brightly striped costumes of a clown can barely compete with the stylish flare of the sea slugs. One species has racing stripes of navy and light blue running down its back, and is ringed by a white and yellow border. There are mottled red, brown, turquoise, and zebra-striped sea slugs, translucent white sea slugs, and at least one species that is white, purple, orange, and blue with yellow polka-dots and purple stripes. Some of the sea slugs wear less brilliant attire, but their coloration so closely matches what they crawl over, it is difficult to distinguish sea slug from seafloor. The colors of the sea slugs are unbelievably flamboyant, but when their strange gill and rhinophore structures are added, these creatures become the epitome of undersea bling.

The small chemosensory tentacles on the sea slug's head, its rhinophores, may be simple rods, tubes, or curly appendages. They can also form spiraling rings, frills, ridges, or tiny branches or resemble oval-shaped feathers. These elaborate shapes are thought to increase the surface area of the rhinophores, thereby enhancing their sensing or "smelling" abilities. The sea slugs' gills can be even stranger, forming frilly edges along their sides, or clusters of quill-like rods running down their backs.

Many sea slugs have a small clump of gills that sit atop their backs, resembling extraneous pieces of algae or curly pieces of lettuce that have been mistakenly attached. Their gills may also be like the ruffles of a fancy shirt or take on the appearance of flattened paddles or delicate lace. They may be similar in color to their bodies or hued in contrast, such as a species whose white body is covered with spiked gills of flaming orange or a pink sea slug that is ornamented with white gills resembling a porcupine's quills. The combinations seem endless. Some sea slugs have a small pocket on the surface of their bodies in which to retract their vulnerable gills should a predator loom near. Others have protective leaves of tissue surrounding their gills. Underwater photographers love the magnificently decorated sea slugs—and, unlike most fish, sea slugs don't dart away while a shot is being framed.

Like most other marine animals, sea slugs spend their days searching for food, avoiding becoming food, and looking to have sex. Using contractions of their muscular foot, they crawl about the seafloor, slithering slowly up and over whatever lies in their path. In coral reefs, where many sea slugs reside, there are numerous obstacles to contend with. A slender strand of algae may necessitate a blind tightrope walk. The steep side of a sponge or face of a coral may require a sea slug to inch its way upward or rappel down. Like a rock climber stretching from one handhold to the next, a sea slug may need to rely on its strong foot to precariously extend its body from one spot to the next. The sea slug's travels are made easier by the secretion of mucus and by thousands of fine, short hair-like appendages that help it to glide over a slick trail of slime. Occasionally, a sea slug ends up crawling across a fish lying in wait at the seafloor. Unless the fish wants to give away its position, it will endure the indignity of the sea slug's travels without flinching. Sea slugs are also known to tailgate; they follow the mucus trail of another sea slug, catch up, and then follow at its tail. Others, to avoid predators, will temporarily become swimmers, using flexible flaps of skin as fins to lift them up and propel them through the water. The Spanish Dancer sea slug is fittingly named for its flamboyant red-and-white ruffled costume and rhythmic body-flexing swimming. Another sea slug resembles a blue, branched twig floating at the surface. A gas bubble in its stomach provides buoyancy, and its blue color comes from its favorite meal—the stinging blue tentacles of the infamous Portuguese man-of-war.

Sea slugs tend to be selective eaters, though their preferences vary between species. Some feed on algae, while others prefer bacteria, coral, sea anemones, small worms, or crustaceans. There are sea slugs that have a fondness for the eggs of other sea slugs or simply for other sea slugs, even of the same species. This can make for some awkward situations. Rejection is one thing, but life is exceptionally tough when your partner attempts to eat you after sex. It is even worse if they try to swallow you whole while having sex! In such circumstances, the larger sea slug usually comes out on top, so to speak, leaving the reproductive organs of its partner for the last, juicy bite. These sea slugs eat by extending their mouths over their prey, much like a snake eating a mouse. Sponge-eating sea slugs secrete digestive enzymes that soften their food into a tasty stew that can be slurped or sucked up. There are also sea slugs that like the rewards of a good wrestling match. Equipped with strong jaws, these sea slugs will attack the tentacles of a tube-dwelling sea anemone. When the anemone withdraws into its tube, the sea slug hangs on and goes in along with it. A wresting match ensues, with the victor, usually the sea slug, coming away fat and happy.

One of the most interesting of the sea slugs' feeding strategies involves an alternative energy source. Algae-eating sea slugs suck out their prey's chloroplasts, which contain chlorophyll and are used for photosynthesis. The kidnapped chloroplasts are stored in the sea slug's body and gills, where they are used as solar cells. The chloroplasts continue to photosynthesize using sunlight, and the sea slug's waste products are used as nutrients, thereby producing oxygen and energy (organic matter), which the sea slug can use to grow. Relatively recent research indicates that solar-powered sea slugs are able to use the captured photosynthetic cells because along with chloroplasts they also transfer genes from the algae. Other sea slugs appear to simply farm and store the algal cells against hard times, consuming them when other food is unavailable. For these sea slugs, at least superficially, they are what they eat, and they very often resemble the algae that they feed on. Carnivorous sea slugs may also become solar-powered by eating other animals, such as corals or anemones, which host symbiotic algae within their tissues. They consume the animal, but save its algal cells to use for solar energy.

As naked, slow-moving snails, the sea slugs appear vulnerable, an easy meal, but for marine organisms that attempt to make a quick snack out of them, it can be a nasty, chewy, and rather bitter lesson. The sea slug's skin is often tough and embedded with glassy spines, and some can secrete toxin or release acid. Sea slugs may obtain their cache of chemical weapons by consuming toxin-laden sponges. Other sea slugs consume organisms such as coral or anemones, which have stinging cells. Somehow they are able to eat the harpoon-like cells of these creatures without triggering their release and then store them for their own use. Fishes that attempt to eat a sea slug have been observed to quickly realize their mistake and spit out the unpalatable snail. The wild coloration of the sea slugs is thought to have evolved as a warning sign to potential consumers—eat me and you will be sorry. There are other organisms, such as worms and sea cucumbers, whose coloration appears to mimic that of sea slugs. Although they are not toxic and may not have the same defenses as the sea slug, using the same coloration is akin to a bluff in a game of poker in which the stakes are extraordinarily high.

Many sea slugs use their coloration and gill structures for camouflage. A bright orange sea slug may spend most of its time on a similarly colored sponge; another might have lacey gills that look just like the delicate branches of the algae or soft coral on which it lives. Some species are able to detach portions of their gills or body should it be necessary, and then regenerate the lost parts. The sea hares have yet another defense mechanism: they can release a smokescreen of purple ink. Even with all the sea slugs' defenses, there are still some marine organisms that are happy to consume a naked snail now and then; these include sea turtles, crabs, sea stars, and some worms or fishes, such as wrasses.

Unfortunately for the sea slug, life is short and time for mating brief. Its life span is often less than one year, although some may live to a ripe old age of four to six. Once they reach sexual maturity, sea slugs often have only a few days or months to live—and to mate. All sea slugs are hermaphroditic; they have the reproductive parts of both female and male. But with the exception of one species that can do it all on its own, it still takes two to tango. Their intimate encounters all occur in the head-to-tail position, because their reproductive organs are always on the right-hand side of their bodies (plate 6A). Each sea slug deposits its sperm into the female duct of its partner; however, for the highly cannibalistic sea slugs, this can truly be a dangerous maneuver and liaison. To prevent disruptions by currents or other disturbances, some species have spines to help lock in place their embrace. It may be a quickie without foreplay or they may nuzzle, bite, and stroke each other prior to mating, sometimes for extended periods. Sea slugs are also known to get it on in groups, forming chains or circles in which three or more organisms mate in unison. The sea slug first in line acts exclusively as a female, while all the others get to play for both teams at once. But the fun never lasts, as they die just days or months later. An abundance of dead sea hares may wash ashore in the aftermath of a food or sex orgy.

The eggs of the sea slugs are laid on the seafloor in masses that could be featured in an art exhibit entitled "Slime in Color." On display would be spiraling ribbons, coiled ruffles, and layered, frilly chains in vivid hues of blue, pink, yellow, or orange; all with one thing in common—a mucus template. The sea slug's brightly colored slime and egg masses resemble decorative cake frosting that has been masterfully squeezed from a tube (plate 6C). The intense coloring of the egg masses is probably a means to warn and ward off potential egg eaters, especially since each one can contain up to 25 million eggs. Most eggs hatch into larvae that temporarily become members of the plankton before settling to a life at the bottom. In a few species, however, a smaller number of hatchlings are produced at a time and they emerge as crawling, miniature versions of the adults.

WHY THEY MATTER

The oceans' snails play a vital role within the marine ecosystem. Pteropods, the potato chips of the sea, are especially important as a link between small plankton and larger marine life within the food web. They provide an energy-laden repast for marine mammals, seabirds, squids, and especially fishes, such as salmon. Conch, cone snails, and sea slugs are less plentiful than pteropods, but they too provide a source of nutrition for organisms with more specialized dining habits and hunting abilities. But the value of snails in the oceans' food web is not limited to just what they eat or what eats them; their waste products are important as well. In the sea, one creature's waste (or egg) is another organism's Happy Meal. And snail poo, in the water or at the seafloor, makes for fine dining for many marine creatures. As for humankind, marine snails have long been of value to people, but their true worth is only beginning to be fully recognized.

Food and Culture

The queen conch's large, whorled, and lustrous pink shell has become symbolic of the sea and its natural beauty. It is prized by tourists, sold with pride in native locales, and frequently used as decoration to depict an association with the oceans. In native cultures, the queen conch's shell has traditionally been used in ceremony and for everyday needs, producing everything from ornaments and trumpets to tools and jewelry. At one time, the conch's shell was used to manufacture fine porcelain and lime. The meat of the queen conch has also long been utilized and prized. It was a principal source of food for early settlers in the tropics and continues to be fished as an important local source of protein, for export, and to showcase regional cuisines. Conch meat is usually served fried, in a ceviche, fritter, or salad, or as part of a chowder or stew. The heavy musculature of the conch foot, however, makes for some rather tough dining if not properly prepared. Soaking in lime juice for several hours will do the trick, or the meat can be ground up or chopped into small pieces. My favorite technique, though, is a less subtle means of tenderization—pounding the heck out of the conch with a hammer before cooking. In the Bahamas, towering piles of old conch shells line the shores, evidence of a once abundant and heavily exploited resource. And in the Florida Keys, natives became known as "conchs" as a testament to the cultural significance of this previously plentiful ocean gastropod.

The oceans' other snail celebrities have yet to catch on as a popular food choice, although nudibranchs are reportedly roasted in the Kuril Islands, and in Fiji, cooked in coconut milk.

Human Health

While the cone snails and sea slugs may not tempt the palate or supply sustenance like the conch, they are providing enormous benefit to humankind in medicine and biomedical research. The cone snail produces a remarkable assortment of powerful toxins, and many of its poisons act selectively on specific target cells or bodily functions. This potency, diversity, and selectivity have made the venom of the deadly cone snail one of the most sought-after compounds for use in the development of new medicines. For centuries, doctors have relied mainly on opiate-derived drugs, such as morphine, to relieve pain. Today, cone snails have provided a new option, one that is especially useful for patients who have built up resistance to other medications or are allergic to opiate-based drugs. In 2004, the Food and Drug Administration approved Ziconotide (also known as Prialt), a painkiller derived from the cone snail's venom. At least seven other pain medications based on the cone snail's toxins are in development and/or clinical trials. Compounds produced by cone snails are also being investigated for other medical applications, such as in the detection of certain types of cancer or in the treatment of strokes, head injuries, and diseases that impact neurologic functioning, such as Alzheimer's, Lou Gehrig's, and Parkinson's. The list of potential applications is long and growing quickly, and includes ailments that involve spasms or convulsions, clinical depression, and cardiac arrhythmias. In a 2003 letter to the journal *Science*, Eric Chivian, Callum Roberts, and Aaron Bernstein suggested that cone snails "may contain the largest and most clinically important pharmacopoeia of any genus in Nature." Of the toxins produced by the cone snail, only a few have been studied in any detail; the possibilities and future potential to help fight disease and improve health care are extraordinary.

Sea slugs are also proving fruitful in relation to human health as models in biomedical research. In 2000, scientist Eric Kandel won a Nobel Prize based in part on his research, which used the sea hare *Aplysia* to study how neurons are able to form and store memories. Who would have equated a slimy, slithering sea slug with a Nobel Prize? The sea hare *Aplysia californica* is a shell-less, mottled, red-brown species found in the shallow coastal waters off California. It has a relatively simple nervous system and a small number of unusually large neurons. Each year, some twenty-five thousand *Aplysia* are successfully bred in the confines of small seawater tanks at the University of Miami's Rosenstiel School of Marine and Atmospheric Science and shipped to researchers around the world. Using *Aplysia*, scientists can examine neurons on a single cell level and are able to investigate the links between individual cells and specific behaviors or functions. As biomedical models they are helping investigators to learn more about nerve-related and brain diseases, such as Alzheimer's and Parkinson's, and to better understand how nerve systems develop. Relatively recent research has also revealed that humans and sea slugs share some similar genes, which may lead to an improved understanding of brain development and related diseases. Given their use of chemical defense, sea slugs are also being probed for potential pharmacological compounds.

Indicators of Change

Like the foraminifera, pteropods can sometimes provide a window into the planet's past. The presence of their shells in seafloor sediments indicates ocean conditions that are or were favorable for their growth. Changes in the abundance of pteropod shells can therefore suggest alterations in water depth, circulation, temperature, or salinity. Because the shells of the pteropods are thin and fragile, they are less likely to be preserved than the more robust tests of other marine organisms. Consequently, they tend to be of greater use in discerning relatively recent histories, meaning conditions within the past two million years. Pteropod shells have been particularly helpful in revealing the details about the recent geologic history of places such as the Mediterranean, the Red Sea, and the North Pacific.

In the oceans today, pteropods are relatively abundant, and because their thin shells are made of aragonite (a relatively unstable form of calcium carbonate), they also provide a means to monitor the ongoing chemical and biological impacts of climate change. The ability of marine organisms to live within the oceans and to build skeletons or shells is strongly dependent on the chemistry of seawater. Since preindustrial times, scientists have measured a slight decrease (0.1 pH unit) in the surface pH of the oceans. It is believed that this small rise in acidity results from the oceans' increased absorption of carbon dioxide as its concentration has risen in the Earth's atmosphere. As climate change continues and more carbon dioxide is emitted into our atmosphere, the oceans' acidity is expected to increase even further. We do not fully understand the impacts this will have on marine life, but early indications are not good and concerns are growing. Because the oceans' acidity affects its carbonate chemistry, the organisms that create shells or skeletons of calcium carbonate are expected to be the most affected. For the pteropods, with particularly thin shells of aragonite, rising ocean acidity could be catastrophic. In the worst-case scenario, pteropods may not be able to precipitate their shells, and if already formed, they could become subject to dissolution. Scientists are now closely monitoring pteropod populations to see what the future will bring. Other research suggests that the oceans' snails are not alone when it comes to the impacts of ocean acidification. Calcification in phytoplankton, zooplankton, and mollusks, such as oysters, as well as in crustaceans and in corals, is also expected to decline or change as the oceans become more acidic.

The oceans' gastropods are diverse and abundant, and an unexpectedly fascinating bunch. They also play a vital role in the oceans, and their shells have become beloved symbols of the sea, delighting serious collectors and the average beachgoer alike. They have long provided humankind with food and tools, and they now hold extraordinary potential for drug discovery and application in biomedical research. And while we have learned much about the marine snails we know and love, there are unquestionably species yet to be found and more societal uses to be discovered. The future holds great promise (and slime) with a sea whose citizenry includes a wondrous assortment of snails.

4 The Riddle of the Reef

What organism in the sea is animal, plant, and mineral all in one? One that also produces slime and has group sex? A hint: these creatures are small, but collectively create some of the largest biologic structures on the planet. The answer? Reef-building corals. They feed on zooplankton with tentacles—the animal. They have symbiotic algae living within their tissues that photosynthesize—the plant. And they build a skeleton of calcium carbonate—the mineral. Reef-building corals are the engineers of a complex undersea topography that is the foundation for one of the most diverse and beautiful ecosystems on Earth. More than four thousand species of fishes have been found on coral reefs, encompassing perhaps 25 percent of all documented types of fishes in the ocean. Add to that, more than eight hundred types of coral along with hundreds of crustaceans, mollusks, sponges, and other invertebrate species. Mix in the microbes, and the diversity housed within a coral reef may be beyond compare. And it all depends on the architecture built by a diminutive but amazing creature—the coral polyp.

THE ANIMAL

Reef-building corals are most commonly found in the relatively warm, shallow, and clear water of the tropics and subtropics. A few corals are solitary, but most live as colonies of small, interconnected animals, called polyps. They are simple creatures, essentially a ring of tentacles surrounding a mouth and stomach. The coral polyp looks very much like its relative the sea anemone, or a bit like an upside-down version of its other cousin, the jellyfish. In a colony, coral polyps are connected to one another by tissue that allows them to share food. Large corals can reach meters in diameter and may contain hundreds of thousands of interconnected polyps. During the day, most polyps are retracted, making corals look more like rocks than a thriving colony of animals. At night, however, they seem to come alive as the polyps extend their tentacles to feed on zooplankton, which are more abundant after dark (plate 7). With their tentacles waving about for food, corals appear like gardens of small wriggling flowers—except these nighttime blooms sting.

When an unwary creature wanders into the arms of a coral, it gets a stinging reception. Like their jellyfish cousins, coral polyps are well armed for capturing prey. Their tentacles are equipped with stinging cells, called nematocysts, each of which contains a small toxin-tipped barb attached to a coiled, spring-like thread. The intensity of a coral's toxin or its sting varies with species, the most potent of which is that of fire coral, which can leave a nasty, burning welt on a person's skin if accidentally touched. After subduing their prey, polyps use their tentacles as feeding utensils, to move food to their mouths and into their stomachs for digestion. They may also use their tentacles to fend off invaders, but corals have another means of defense—they can reach out to not-so-nicely touch their neighbors.

On coral reefs, real estate is at a premium. Corals must compete with one another and other organisms, such as sponges and algae, for a place to live and to grow. They must ward off faster-growing species to avoid being smothered or left situated in the shade. If another organism gets a little too close for comfort, many of the hard corals have a "gut reaction"—they extend stringy strands from their stomachs to the site of the incursion. It is an "out-of-stomach" experience in which the filaments release digestive enzymes that eat away at the invader. A "kill zone" of several centimeters may be set up around an aggressive coral in which intruders are attacked or space is cleared for expansion. Some coral species are more aggressive than others in using their mesenterial filaments. The massive head-type corals are the most hostile, but they are also relatively slow-growing as compared to platy or branching species. Their aggressive behavior may enable them, as slower-growing corals, to coexist with faster-growing species. Corals may also set up a nighttime perimeter, guarded by special tentacles. These "sweeper" tentacles tend to be three to five times the length of a coral's regular tentacles, have enlarged stinging cells, and come out principally at night. They are long-arm protectors that wave back and forth at the edge of a coral colony to provide a stinging rebuff when a sponge or other intruder gets in range. Sweeper tentacles may also be used to ward off an adjacent coral's mesenterial filaments, somewhat like an adult holding a child at arm's length as he or she tries to throw a punch. On a reef, the war over space is never-ending, and for corals, it is a battle that requires life- and tentacle-long vigilance.

As sedentary creatures with a colony of mouths that open principally upward, corals also need to contend with potential smothering or choking by sediment and particulate material raining down from above. This is where slime comes in, coral mucus to be precise. Corals have an overcoat of mucus, which entraps particles that fall on them. By secreting additional mucus, they can then slough off a layer of slime along with any goo-ensnared particles. A coral's mucus may also be used to capture bits of food or as protection if it becomes exposed to air at extremely low tides. Bacterial communities within coral mucus appear to play an important role in their ability to fight off disease. And when corals are physically disturbed, they sometimes exude profuse amounts of slime in stringy strands of undersea goo, known as "coral snot."

And now, for some coral sex. Reef-building corals can reproduce in a variety of ways. A colony of coral polyps may grow by simple asexual budding, one polyp dividing into two or a new polyp forming in a space between two adjacent polyps. Polyps may also detach or bail out from their parent colonies, settle to the seafloor, and begin growing as new coral outposts. Corals that are branching or platy are also able to grow through fragmentation. Pieces broken off by storms, clumsy predators, or some other physical disturbance can, if conditions are right, attach to the seafloor and begin to grow. When a coral reproduces asexually, the new polyps are genetically identical to the parent—they are coral clones.

Sexually reproducing corals are mostly broadcasters, which means they release packets of eggs and/or sperm into the water. The corals' eggs have a high fat content, and once released, buoyancy causes them to float to the surface. The eggs can act as energy-saving lifejackets for the corals' sperm: when eggs and sperm are packaged together, both are carried upward. Some corals release eggs and sperm separately. Broadcasting increases dispersal and genetic mixing, but is only effective if sex is a group affair. Corals of the same species need to release their eggs and sperm at about the same time to improve the likelihood that fertilization will occur and viable offspring will be produced.

Synchronous spawning (aka group sex) in reef corals is an annual affair that can last for hours; it occurs regionally on reefs at about the same time and day each year. In South Florida, the action usually takes place around the full moon in August; in Okinawa, Japan, around the full moons in June and May; in Guam, at about the full moon in July; and in Australia, around the full moon in November. Some coral species may release their egg bundles early in the evening, while others do it later at night. Although no one knows precisely how corals synchronize the release of their bundles of joy, the factors involved may include warm water temperatures, moonlight, sunlight, and possibly the tidal cycle or chemical cues from other corals. When corals engage in group sex, it creates a colorful, upside-down undersea snowstorm that instigates a nighttime feeding frenzy. Creatures such as worms, fishes, and even whale sharks, come to partake of the rising flurry. If the corals' eggs and sperm manage to successfully pass through this gauntlet of hungry mouths and fertilization occurs, embryos form and develop into small, floating larvae.

Young reef corals may float for days to weeks before finding just the right place to settle onto the seafloor and morph into juveniles. They need an appropriate substrate with favorable salinity, temperature, and stability, along with sunlight, and some require specific algae to provide chemical cues. As the baby corals search for a suitable landing site, they must contend with more of the oceans' starving creatures, strong currents, storms, and sometimes just bad luck. Broadcasting corals may release hundreds to thousands of eggs, but only a few, if any, larvae may make it through the oceans' obstacle course. Survivors may grow close to parent corals or be carried far away, depending on the flow conditions during and after spawning.

Some corals brood their young; for them, satisfaction has to come from within, as eggs are self-fertilized. Larger, fewer, more developed offspring are released into the water, where they typically settle relatively close to the parent coral. This illustrates two of the contrasting and common reproductive strategies in the oceans: (1) the release of many relatively undeveloped eggs or young in the hope that a few make it, and (2) investing more energy into fewer young that are born better developed and more able to survive.

THE PLANT

In a true partnership for the ages, all reef-building corals in relatively shallow water have a symbiotic relationship with zooxanthellae (pronounced zō-ə-zan-´the-lə), which are small, one-celled algae or dinoflagellates. Millions of brown or yellow-brown zooxanthellae may reside within a coral's tissues, where they get a protective home along with several of the essential substances needed for photosynthesis. Corals provide the algae with carbon dioxide through respiration, and nutrients from their wastes. In return, the coral host gets an internal manufacturing plant that produces oxygen and organic carbon (food) through photosynthesis, and in the process, recycles its garbage. The use of phosphorous, one of the coral's waste products, by the zooxanthellae may be critical to a coral's ability to produce a limestone skeleton, because phosphate is believed to inhibit the precipitation of calcium carbonate. Corals without algal symbionts do not typically build massive limestone reefs; they get their nutrition solely from feeding on zooplankton and can live in darker, deeper, and colder waters.

The corals' symbiotic relationship with zooxanthellae provides a competitive advantage in the nutrient-poor environment in which reefs tend to grow. The coral and zooxanthellae rapport, however, is not perfect. Under stress, a coral may "bleach," or release its colored algal symbionts, leaving its tissues transparent and the underlying white skeleton visible. Some bleached corals can continue to live without their zooxanthellae, by using their tentacles to feed on zooplankton. These corals may eventually bring zooxanthellae back on board and recover. In severe or extended cases of stress, corals die from the loss of their algal associates. Prolonged high seawater temperatures, unusually cold conditions, or high ultraviolet radiation can cause corals to bleach. Recent research suggests that some corals may be able to switch their algal partners after bleaching and that certain zooxanthellae are more resistant to heat than others, two factors that could potentially help some corals adapt to and survive climate change.

THE MINERAL

Most of us think of a coral reef as a large living structure. But even under the best of circumstances, most of a reef is dead. Beneath a thin veneer of living tissue at the surface lies the bulk of a reef, an accretion of limestone that typically ranges from a few to thousands of meters in thickness. Within this accumulated framework, holes and cracks are filled by sediment, and other reef creatures add cementing pavements and tubes of calcium carbonate. Incredibly though, the bulk of a reef's massive accumulation of rock is created by corals and their simple, yet industrious polyps.

Within each colony, polyps sit in little skeletal cups of calcium carbonate or limestone. As the coral grows, its skeleton accretes, and its polyps essentially get jacked up over time. Corals annually produce two skeletal layers, much like tree rings. A thin, high-density band is formed in late summer, and a thick, low-density band is produced throughout the rest of the year. Branching corals tend to grow relatively fast, some 20 to 30 centimeters (8 to 11 inches) per year, while mound and head corals typically grow more slowly, about 5 to 10 millimeters (less than half an inch) per year. Large, individual corals living in the ocean today may be quite old, hundreds to thousands of years old. The reefs they reside in may be even older; many could have started forming thousands of years ago.

While the bulk of a coral reef's limestone lies below, it is at the surface that its complexity is at a peak, and it is here that the overlapping hodge-podge of shape and form create an astonishing underwater landscape and what is arguably the greatest city of life under the sea. Corals are the foundation and condos of this metropolis, and they come in a wide variety of architectural designs. There are purple and tan finger corals that form individual mounds or rolling, colonial hills. Platy brown or spiraling green corals produce the reef's flat plains and low-rises, while towering tan pillar corals are its skyscrapers. Variously colored branching forms may sit scattered about or cover the reef in a dense horizontal sprawl. Rounded brain and mound corals are the mountains of this seascape, and spiked or wavy tabletops add structural variety. The elaborate architecture of the coral reef rivals that of the world's rainforests and similarly supports a great diversity and high density of residents, including many interesting and rather bizarre characters.

RESIDENTIAL LIFE

Life abounds on a coral reef. Thousands, if not millions, of organisms live within its holes and on its surfaces, and swim above and throughout its framework. A brief look at a few of the reef's most fascinating inhabitants reveals much about this vibrant city under the sea and the lifestyles of its diverse residents. Undoubtedly, the most conspicuous creatures are the fishes, the tenants of scale and tail.

Reef fishes form a collection of odd characters whose shape, size, and color vary along with their social skills and sexual proclivities. Some fishes are gregarious, forming roving bands that ramble about the reef throughout the day. Others prefer a more solitary life or the company of a mate. Many of the reef's fishes are most active during the day, while others favor the nightlife or prefer a time of shadows, just before dawn or dusk. This so-called changeover time is when many of the reef's top predators take to the hunt.

On the coral reef there are meat eaters, vegetarians, and those whose dining preferences are a bit more versatile or opportunistic. Some reef fishes set up territories where they patrol to protect their food, the best hiding holes, and nesting sites. Others create and aggressively protect small farms of algae so that fresh food is always on hand. And there are the wanderers, fishes that roam the reef but generally stay within a specific home range.

Of the reef's large carnivores, perhaps one of the best recognized (and feared) is the barracuda—second only to sharks. The barracuda's fearsome reputation may arise from its sleek, silvery, missile-like body, its gaping mouth packed with large, razor-sharp spikes, or maybe just its all-too curious nature, as barracuda have an unnerving tendency to hover or follow closely behind divers. But barracuda rarely attack humans, and if they do, it is typically a case of mistaken identity, especially in murky water. Humans are truly not good fish food. The barracudas' propensity as a stalker is not an attempt to peruse the menu; rather, they are strongly territorial and quick to investigate invaders within their domain, albeit fish or human.

Like the barracuda, the moray eel is one of the reef's reputed bad guys—probably due to its big, yawning mouth that is also full of razor sharp teeth (plate 8). Yet, like the barracuda, its reputation is ill-founded. Moray eels are for the most part reclusive, rather shy creatures. At night they hunt for prey, such as fishes, squids, octopuses, and crabs, but much of the time they hang out in their holes, just watching the reef's activity. People get a little freaked out (some more than others) by the moray's habit of opening and closing its mouth, making it seem at the ready to strike, chomping at the bit. But this is not an aggressive behavior; it is how they move water in and over their gills—they are simply breathing. Moray eels are also friends of the slime. Their long, scaleless bodies are covered with a protective coating of mucus.

Groupers, grunts, jacks, and snappers are also among the reef's most common predatory fishes; their diet of choice is mainly small fishes and various crustaceans. They are all pretty much shaped like a "typical" fish, though each has its own slight variation in form. Grunts are known for their namesake vocalizations that are created as they grind their teeth and the sound is amplified by their swim bladders, which are internal gas-filled organs that help fishes to control their buoyancy. The groupers, especially the giant goliath grouper, emit an even more startling noise. Through contractions of a muscle that is connected to its vertebrae and swim bladder, the goliath grouper creates a deep, thunderous boom or bark. It is thought that the grouper uses its bark as a warning or as a means to communicate to its fellow fish, particularly during aggregations for spawning. And when it comes to sex, groupers like a party. In fact, not only do groupers aggregate for spawning, they also have special places for the festivity. During specific times of the year, they travel from far and near to specific rendezvous spots to bring forth a new generation. Groupers have been described as the puppy dogs of the ocean and will sometimes approach a diver wide-eyed and curious. Not a good thing if the diver has a spear in hand, but for the observer or photographer, such encounters are magical, allowing a mutual inspection that brings two very different creatures of the Earth seeing eye to eye.

When it comes to the reefs' more elegant residents, the spotted eagle ray wins hands down. Like a winged bird in flight, this cartilaginous fish gracefully flies through the sea seemingly without effort. The spotted eagle ray has a diamond-shaped body tipped by a long, whip-like tail. Its mouth, located on its underside, sits just behind a short, pointed snout, and is perfect for scooping up and munching on small fishes, worms, and crustaceans found at the seafloor. The spotted eagle ray gets its name from the white, blotchy polka-dots that adorn its typically dark purple or yellowish back. Underneath, the ray is milky white. Being dark on top and light on the bottom is a coloration called countershading, a common feature among organisms that spend a good part of their lives swimming in the oceans' sunlit waters. Countershading provides excellent camouflage from predators looking up into the sunlight from below and those peering downward into the oceans' dark depths. It also makes fish hard to see from the side, and when combined with their underlying shadow, allows organisms to blend in with the sunlight filtering through the sea's blue waters. The spotted eagle rays are well-dressed, unaggressive, and truly magnificent creatures. They also have a habit of leaping out of the water with great flair, possibly to clean off parasites, when being chased or as some form of communication (we don't really know why).

Among the reef's herbivorous fishes, there is also a definite standout. It is a common creature that has big buckteeth, can be wildly hued, has the ability to change sex on request, and helps to create the fine white sand so symbolic of the tropics. Named for their fused teeth or "beak" and wonderfully bright colors, parrotfishes are hard to miss. Large species can reach up to ⅔ of a meter (about 2 feet) in length as adults. They have visibly distinct scales and a thick, elongated oval body that tapers to a forked or fan-like tail. Some of the most beautiful parrotfish are an astonishing shade of turquoise that is embellished with stripes of white, pale pink, dusky purple, or bright yellow. As a telltale sign of species, however, color can be deceptive, varying within a species and changing with age, sex, and during social interactions. Scars on the reef's surfaces are evidence of the parrotfishes' propensity to scrape, chomp, and bite off algae. Sometimes they will also nip at sponges, coral polyps, or the blades of seagrass, but their main diet is algae. The strength of their bite causes the parrotfishes to ingest significant quantities of the underlying limestone; they are not the most delicate of diners. As extremely proficient grazers, parrotfishes are also another of the oceans' prodigious poopers, frequently releasing clouds of excrement that help to recycle nutrients and produce the tropics' lovely, fine, white sand (maybe not the thing to think about while sipping piña coladas on the beach during your next tropical vacation).

Parrotfishes also have an exemplary sex life. Individuals often live an alternative, transgender lifestyle, beginning as females and changing into males at a certain size or age or if the social situation requires it. If a harem of parrotfish loses its dominant male, no problem—within ten days, one of the older, larger females becomes a male. It is a handy way to reduce the likelihood that an outsider will stage a coup and take over the group. Some parrotfish never change gender, but once a female goes male, there is no going back; the gender swap is irreversible. Females that change sex and become dominant males may become "supermales," larger, more aggressive, and particularly colorful. Research suggests that hormones drive their gender change. When it comes to having sex, some parrotfishes spawn in pairs, while others prefer more action and do it as a group. To put the females in the right mood, the dominant males put on an enticing show for potential mates, folding their fins, arching their tails, or exhibiting jerky movements—it is not an especially graceful seduction. They may also chase the females or make rapid vertical swims. As seems to happen quite often in the oceans, while the dominant males are showing off and engrossed in their display, "sneaker" males may dart in for a quick hookup. The actual act of spawning occurs when a female parrotfish releases her eggs into the water and a male parrotfish discharges sperm so that the two mix.

Some parrotfishes also have an odd sleeping arrangement. At night, they settle into the safety of holes and crevices in the reef for a snooze, wrapped in a security blanket of slime (plate 9). These parrotfishes spin a protective cocoon of mucus, which is thought to be toxic or mask their scent from predators, such as sharks or moray eels.

These next reef fishes do not have large teeth—thank goodness—or spin a poisonous blanket of slime, but they can be hostile and sneaky; they are the oceans' version of small, ankle-biting dogs. Damselfishes are the reef's best farmers and some of the most aggressive fishes in the sea. Luckily, they rarely grow larger than about 15 centimeters (6 inches) long. Some damselfish species feed on plankton in the water just above the reef, but many tend to small crops of filamentous algae, which they protect with great fervor. These plucky creatures will not hesitate to attack any organism of any size that strays into their pastures.

One particularly aggressive type of damselfish is the sergeant major. They are named for the five black, vertical stripes running down their small yellow and white bodies. The sergeant major is opportunistic about its diet, eating whatever is available: algae, plankton, small invertebrates in the sand, or even a fish now and then. When brooding their eggs, as all damselfishes do, they are at their fiercest. After a male sergeant major has courted and successfully wooed a female, she lays her eggs in a nest he has prepared and shown off. The male then stands guard, wary of potential egg eaters. Intruders are warned off with a flash of darkened color as he passes back and forth, agitated, over the nest. If needed, however, the egg guarder will not hesitate to go on the offensive, in for a strike. Sergeant majors seem to be especially clever about it as well, biting from behind, when their targets are not looking. Though blood is rarely drawn, it can feel like a strong knock on the head or sharp pinch to the derriere when attacked by a feisty sergeant major—I know.

There is at least one fish on the reef whose big mouth is used for more than just feeding or biting; it is a tool for home improvement and an incubator for its eggs. The relatively small jawfish has an elongated and narrow body, a blunt head, large, bulging eyes, wide puffy lips, and a really big mouth. Jawfishes use their mouths as front-end loaders to excavate burrows in the reef's sandy areas and then spend much of their time picking up and removing sand and rocks from their homes—maintenance is a mouthful. Male jawfishes also use their mouths as incubators for their eggs (plate 10). After an elaborate courtship dance, a little private burrow time, and the laying of the female's eggs, the male becomes the parental caretaker. He picks up the couple's eggs and holds them gingerly within his mouth. For some five to seven days, with bulging cheeks, he will care for and incubate the eggs, periodically spitting them out and then sucking them in to aerate, clean, and rotate his charges, helping them all to develop. Once hatched, the baby jawfish leave his care and join the ranks of the plankton before settling to a life in the sand.

Going on appearance alone, some of the reef's oddest fishes are those that are experts in disguise. Among the reef's stealthiest are the scorpion- and trumpetfishes. They know just how to use their looks to get what they want. The mottled coloration and fleshy skin appendages of the scorpion- fishes allow them to blend in amid the reef's algae-covered rocks. Utterly motionless, a scorpionfish's body, broad mouth, large eyes, and fins are difficult to distinguish from algae and rock. Scorpionfishes stand out only when swimming from one spot to the next as they extend their magnifi- cent pectoral fins, which resemble multicolored Japanese fans. Scorpion- fishes also have sharp, venomous dorsal spines that are used in defense. They are remarkable creatures to observe if you can actually find one. The trumpetfishes are a bit easier to spot. They have long, ruler-like bodies and heads, and mouths that when expanded look like trumpets. This fish uses its unusual body shape to hide, hanging upside down and motionless, of- ten within the similarly shaped arms of a sea whip. The trumpetfishes are also known to shadow hunt, swimming closely alongside other fish, es- sentially in their shadow, to ambush prey.

Many reef fishes do not possess large teeth, or burrow, or blend in well with their surroundings. But with an abundance of hungry predators plying the world's reefs, these fishes have had to evolve other means of deterring diners. The puffer and porcupine fishes can suck in water and inflate, making their balloon-like bodies appear much larger. Some of the pufferfishes also contain a deadly toxin, like the infamous Japanese fugu. The trunkfish seems a vulnerable creature, resembling a small, black- and-white polka-dotted box that maneuvers slowly about with tiny fins; however, like some of its relatives, this fish can secrete poison from its skin. One of the loveliest creatures on the reef is the Caribbean's queen triggerfish, wreathed in pastel shades of blue, purple, yellow, and green with delicate, wispy dorsal fins and dainty streamers running off its forked tail. But don't try to swallow this beauty—at its head is a dorsal spine that can be raised and locked in place. In the Pacific, the clown triggerfish is comically dressed with a polka-dot belly and striped head; but it too is similarly armed and "trigger" ready. A wide variety of surgeonfishes reside within the reef as well, and each has two scalpel-like spines just in front of its tail. If provoked, the spines jackknife open. The reef is resplendent with savory fishes, but many have some rather unsavory attributes.

Some of the reef's fishes have more passive means of defense. The flat, triangular shape of the angelfishes allows them to keep a narrow profile and makes them difficult to take a bite out of. The four-eye butterflyfish has a fake eye near its tail as a decoy for predators that may be aiming for its head. It is easier to escape when attacked at the back end. The film *Finding Nemo* brought the reefs' clown- and anemonefishes wide fame, and they have since become the darlings of aquariums. These fishes have a very different approach to avoid being consumed. Found in the Indian and Pacific Oceans, they are small—okay, cute—fishes that may be orange or yellow with white vertical stripes. They spend much of their lives hovering within the stinging arms of sea anemones. They get protection, while their anemone hosts get a cleaning service and security guard. What protects these fishes from the anemones' sting? Why, slime of course! They sport a protective wetsuit of mucus.

One of the most fascinating sights on a coral reef is when a fish stops in for a clean and shine. The client fish, such as a grouper, parrotfish or moray eel, enters an established area or cleaning station, hovers motionless, and opens its mouth wide. Some will pose in a head-down position, while others may flash a color change. Once the client is in place, a small fish or shrimp approaches and begins picking off parasites, injured skin, or bits of mucus. Cleaners will even enter a client's mouth to ply their trade (plate 11). It is a neutral zone in which the cleaners get a free meal and the fishes get a good scrub. There are those on the reef, however, that will take advantage of the peace treaty between predator and prey. The sabretooth blenny is a cleaning station con artist that resembles and behaves like a cleaner wrasse. After an unsuspecting fish settles in for a wash, the sabretooth blenny stealthily moves in as if to clean, but then, without warning, sinks its enlarged canine teeth into its unwary victim, stealing a quick and luscious bite to eat.

Among the reefs' fishes are also the tangs resilient in bright blue or yellow, the funny-snouted unicornfish, the artistically patterned filefishes and cowfishes, and the magnificent spotted drum with a tall crown and flowing robe of black and white. The list is long and varied, and it is tempting to go on, but there are other stories to be told of creatures less well known, yet no less intriguing.

OTHER CITY DWELLERS

Corals and fishes are not the only interesting inhabitants of the reef. Other common, but in no way ordinary, residents of this undersea city are the algae—and they are definitely not the green slimes of your aquarium or backyard pool. They are an odd collection of seaweeds that come in a diverse array of colors and shapes, and have some surprising strategies to propagate and avoid being salad for the reef's grazers.

The green alga *Halimeda* is an excellent example, often growing in great abundance on reefs. It is a conspicuous species with an underlying skeleton of linked calcium carbonate plates, which often look like cornflakes frosted in green. *Halimeda* may form densely packed clusters of these verdant cornflakes or branching chains. When the plant dies, its skeletal flakes fall to the seafloor, becoming sand. In fact, *Halimeda* is one of the top manufacturers of sediment in the tropics; mass-producing much of the sand that accumulates in the reef or washes ashore onto beaches.

For the reef's herbivores, however, *Halimeda* is a pretty poor food choice, much like munching on pebbles covered with just a thin coating of delectable tissue. The exception is when it is first grown; forty-eight hours are needed for each segment to form its underlying and crunchy skeletal plate. *Halimeda*, though, has evolved means to thwart would-be consumers of its vulnerable and tasty new tissue. Growth occurs principally at night, when most herbivores are inactive, and its youngest tissues contain a chemical that acts like an herbivore repellent, should a fish try for a little midnight snack. One of the reef's small crustaceans, an amphipod, takes clever advantage of this toxic new growth. During the night it wraps itself up in a freshly grown piece of *Halimeda*, creating a home with an in-house repellent against predators; expert Mark Hay from the Georgia Institute of Technology likens it to the creation of a toxic burrito.

It may surprise you to learn that this seaweed has sex, and like many of the oceans' other residents, it prefers to do it as a group. *Halimeda*, along with several of the reef's other green calcareous algae, can reproduce sexually by releasing male and female gametes (sort of like eggs and sperm in animals, though both sexes are motile) into the water. Ken Clifton of Lewis & Clark College discovered that for about twenty-four hours before releasing its gametes, the alga prepares by moving all its live tissue to its outer tips and converting it into reproductive cells. When the morning of its special day arrives, the *Halimeda* releases its swimmers in thick green clouds or slimy streams. The male gametes are released first, presumably to improve the chance of fertilization, as the females must then swim upward through a cloud of males. Fertilized eggs sink to the seafloor to begin growing anew. For the "parent" *Halimeda* plant it will have been an all-out effort to procreate, because after its sexual release, it dies.

Among the reef's other green algal oddities is a species resembling a thin vine adorned by clusters of tiny balls, like miniature grapes, or small feathers, and another that stands upright resembling a shaving or cosmetic brush. One of the most bizarre seaweeds is the "Sailor's eyeball," a small shiny, round orb of an alga. Strange types of brown algae also abound. The brown algae *Dictyota* can, at times, grow like an undersea weed on the reef, with stubby, thin, and flat fronds that seem to cover nearly every available surface. Sometimes, when the light is just right, hues of purple and iridescent blue appear almost magically among its dull, tan branches. *Dictyota* and some of the green algae, including *Halimeda*, can also spread incredibly fast through fragmentation and asexual reproduction. Scientist Steven Miller of the University of North Carolina at Wilmington likens it to leaves falling off a shrub, being blown by the wind, and then settling on the ground, where within days, they begin to grow as a new bush. Divers, boats, fishes that are sloppy eaters, or storms can fragment these types of algae and cause them to spread like wildfire.

Red algae are also abundant on coral reefs. The most common forms create small bushes of sticks or beautiful pink and purple pavements that help to harden and bind the reef's framework. Coralline algae, a form of red algae, are believed to provide the chemical cue that some coral larvae need to settle down and morph into juveniles.

Also making the list of the reef's most fascinating residents are the sponges, simple, but ancient organisms that evolved some 600 million years ago. One of the reef's most stunning residents is in fact a sponge; it forms an iridescent purple-blue vase ornately decorated with delicate ridges and valleys. Some sponge species, however, are not quite as lovely, like the brown, shiny blob variety, appropriately named the chicken liver sponge. The most noticeable sponges are probably the large tube, pipe, and barrel varieties that come in an assortment of colors, including purple, yellow, and brown. Sponges may also grow as brightly hued, irregular encrustations or in elongated ropey forms. The largest is the giant, brown barrel sponge, which can grow up to 1.8 meters (6 feet) tall.

Attached to the seafloor, sponges appear to be idle creatures. Yet they are surprisingly active organisms that draw seawater in through small perforations throughout their bodies to filter out oxygen and particulate organic matter for food. They then pump the seawater out through one or more larger holes. And some sponges are serious pumpers. A team of researchers led by Chris Martens and Niels Lindquist at the University of North Carolina at Chapel Hill discovered that a large barrel sponge can pump a volume of water equivalent to that in an Olympic-size swimming pool in just one day!

Sponges often contain needle-like spicules of silica or calcium carbonate for structural support and toxins that can be used as chemical weapons against potential sponge eaters. Some sponges can even secrete acid to drill or bore their way into the reef. Need I say, it is best not to touch sponges on a reef, especially the aptly named "fire " or "touch-me-not" sponges. Relatively recent research has also revealed that many sponges are jam-packed with microbes, which seem to play an important, but poorly understood, role in their biology.

Occasionally, reef sponges resemble erupting volcanoes. Emanating from their tops are smoke-like plumes that are actually clouds of sponge sperm being broadcast into the water. Other sponges suck in the little swimmers, which are either eaten (not the same species) or used to internally fertilize their eggs (the same species). Sponge larvae are later broadcast from the parent sponge into the surrounding seawater. Some sponges are also able to reproduce by simple asexual budding, aka cloning.

Other marine life make good use of the reef's sponges. Organisms such as brittle stars, shrimp, and an assortment of worms take advantage of the sponges' architecture by using them as prefabricated homes. The decorator crab has another use appropriate to its name; it captures a piece of sponge and carts it around on its back as camouflage.

A group of unexpectedly colorful characters on the reef are the worms—yes, worms. The flatworms of the coral reef resemble small, moving, multihued pancakes that can be as flamboyant as the sea slugs, the neighbors they mimic. Tube-dwelling worms are an even stranger bunch. They live within a small tube in the reef and extend their tentacles out into the water to act both as gills and as a net for capturing particulate food material. The surprising resemblance of their tentacles to everyday items has led to some especially fitting names, such as the feather duster, spaghetti, or Christmas tree worms. The bristle- or fireworm is also a handsome creature, but one that should be kept at arm's length. It is a slender, slightly flat, segmented worm with furry white and sometimes red tufts lining its sides. They may look soft and inviting to the touch, but their tufts are actually made of small, hollow, and very sharp spines. When threatened, the fireworm flares out these bristles, which upon contact penetrate the flesh, inject a mild venom, and break off. An unlucky run-in with a fireworm produces a painful, burning welt, hence the name. Fireworms are usually about 7 to 10 centimeters (about 2 to 4 inches) in length, though they can grow larger. They feed mostly at night and have an insatiable appetite for coral polyps, anemones, and small crustaceans.

Among the other interesting residents of the reef are some notable shellfish, such as the thorny oysters, the promiscuous queen conch, and of course, the beautiful but deadly cone snails. But possibly the largest and most famous, or maybe infamous, of the bunch is the giant clam. Found on reefs in the Indo-Pacific region, the giant clam is one big shellfish, with the largest species, *Tridacna gigas*, growing up to 1.5 meters (5 feet) across and reportedly living for up to two hundred years. One massive specimen weighed in at more than 226 kilograms (500 pounds)—try eating that on the half shell. Due to its massive gaping shell and some rather dramatic, but dubious tales, the giant clam has gained a nefarious reputation and been dubbed the "killer clam." Authenticated incidents in which a giant clam has actually entrapped and drowned a diver, however, are notably lacking. The notoriety of this giant mollusk may have been fueled by one particularly lurid tale that includes at least one murder and an enormous pearl worth millions. It is the story of the "Pearl of Allah."

The Pearl of Allah is said to be the largest clam pearl in the world; it has also been called the ugliest. It is a matte-white wrinkled mass, 23 centimeters (9 inches) long and weighing more than 6 kilograms (14 pounds). In a 1939 article in *Natural History*, archeologist Wilburn Cobb wrote that the pearl was presented to him on the island of Palawan in the Philippines after he saved the life of the chief's son from malaria. The local chief had supposedly obtained the pearl from a giant clam after it had caused the untimely demise of a young, native diver (really?). The chief had then, reportedly, christened the large gem the Pearl of Allah because its surface reminded him of a turbaned Muhammad.

The gigantic pearl was subsequently put on exhibit at Ripley's Believe It or Not in New York City and appraised at a value of $3.5 million. It then mysteriously disappeared, only to reappear some thirty years later up for sale and with a new tale to be told. Cobb claimed that a man from China had visited Ripley's museum and asserted that the Pearl of Allah was actually an ancient Chinese gem, the lost Pearl of Lao Tzu. After making an unsuccessful offer for the pearl, the man reportedly disappeared and was never heard from again (hmmmmm). Upon Cobb's death in 1980, the pearl was sold for just two hundred thousand dollars. The new owners of the giant but gnarly gem put it up for collateral on a loan and were subsequently taken to court for collection. The saga continued when the children of one of the next owners sought the Pearl of Allah as an award in a wrongful death suit after their father was indicted for the murder of their mother. The pearl was finally ordered sold in 2007 at an appraised value of $32.4 million. This extraordinarily high appraisal value was suspect, however, due to the pearl's questionable origin and history, and its unknown age. The Pearl of Allah remains mysterious and its worth debatable, but one thing is now clear: the giant clam is no killer.

With its enormous waved shell sitting agape, the giant clam certainly seems ready to clamp shut on the unwary, but this resident of the reef is one of the sea's gentle giants, a filter-feeder of plankton and particulate matter. There are some seven species of giant clam, all of which obtain food as water is drawn in through a tube or siphon, filtered, and then expelled through an exhalant opening. Giant clams also have zooxanthellae living within their bodies, which add in nutrition and help them to grow to their unusually large size. Juvenile giant clams initially ingest the zooxanthellae, then move the algae from their stomachs to their outer body walls or mantles, which appear like colorful, fleshy cushions in the gap between the animals' two shells. In some giant clams the mantle is spectacularly colored golden brown, yellow, or green, and speckled with hundreds of iridescent blue or purple spots. For self-protection, the giant clam has primitive eye-spots that are like light-sensitive burglar alarms: when triggered, they cause its shell to close, thereby protecting the vulnerable creature inside. This megamollusk also has iridophores, or clusters of pigment cells, that give rise to a blotchy or striped pattern in its mantle. The iridophores help to redirect light to or away from the zooxanthellae within the giant clam's tissues, sort of like tiny, rotating solar reflectors.

With thousands of diverse residents, the coral reef has a seemingly endless supply of weird and wonderful creatures. There are also sea urchins all dressed in spikes, sea fans that wave elegantly back and forth, and stunning soft corals, which resemble bushes of cotton awash in an assortment of vibrant colors. Shrimps, crabs, and lobsters ply the reef's crevices, while sea anemones and sea squirts attach to its surfaces. There are sea cucumbers, sea lilies, sea baskets, and sea turtles to be found among the corals, as well as sea stars, squids, and octopuses. This city's residents are an amazing lot and there are far too many to describe here. Along with a huge diversity of organisms, amidst the coral reef there are also a seemingly endless number of interesting behaviors, lifestyles, and interactions occurring. No matter how often you visit or how long you stay, there is always more to experience and to see.

WHY THEY MATTER

Coral reefs provide invaluable housing and food to the creatures of the sea, but they are also of great worth to human society. In terms of economic value, coral reefs are not just big business; they are the equivalent of an international megamoney conglomerate. It is conservatively estimated that globally, coral reefs provide seafood and services worth more than $375 billion dollars each year. In 2010, only one company on *Fortune Magazine*'s top 500 of the world's largest corporations had higher annual revenues. Each year, coral reefs provide more worth than the top oil companies, automobile makers, banks, communication giants, or investment firms. The net worth of the wealthiest individuals on the planet is about one tenth of what coral reefs are worth each year. And that does not even take into account the aesthetic value of coral reefs, their ability to educate and inspire, or their worth as repositories of potential drug discoveries. From grouper sandwiches to cancer-fighting drugs and millions of jobs, coral reefs are global providers of goods and services annually worth hundreds of billons of dollars.

Food

On the coral reef, sharks and barracudas are impressive hunters, but they cannot hope to compete for the title of top predator if humans are among the challengers. People across the world consume reef species in great quantity. For some diners, a fresh grouper fillet or steamed lobster from tropical waters is a treat to be savored on a night out or while on vacation. In developing countries, however, coral reef fishes are about one quarter of the total fish catch, and in Southeast Asia they provide food for an estimated one billion people. For many communities, fishing is not only a way of life, but also a necessity for survival. It is not just the finfish or crustaceans of a reef that are consumed—around the world people dine on the resident shellfish, sea cucumbers, sea urchins, and more.

Recreation, Tourism, and Trade

The marine life associated with coral reefs provide people across the globe with a valuable commodity for trade and sale, as well as the foundation for an enormous recreation and tourism industry. People will travel vast distances and spend lots of hard-earned cash to explore a coral reef or to enjoy a few days relaxing on the soft white sand of nearby beaches. Along with scuba diving and snorkeling, coral reefs provide a base for recreational fishing and boating activities and all the associated support services, such as restaurants, marinas, hotels, airports, and souvenir shops. In both South Florida and Hawaii, reef-based tourism and recreational activities annually generate more than a billion dollars for the regional economies. And what about the intrinsic worth of wonder or relaxation? Can we put a quantitative value on that? Where coral reefs create a good wave break for surfing, how is that priced?

Over one million people visit aquariums each year, and some of the most popular exhibits are based on coral reefs. Our love for these amazing ecosystems may be best illustrated by the fact that millions of people try to re-create them in their own homes or businesses. In the United States alone, some six hundred thousand people keep aquariums. The aquarium industry is a multibillion-dollar business, much of which is still based on collecting wild fishes from coral reefs. More than forty-five countries supply over 10 million aquarium fishes each year, with damselfishes, anemonefishes, and angelfishes making up some 50 percent of the volume. Invertebrates, such as the blue sea star and anemones, are also traded. In some countries the ornamental fish trade provides one of the few options for employment. Corals may also be collected, trade, and sold for aquarium use, decorations, and jewelry. In some regions, thankfully, it is now illegal to take coral from the sea or to export it for any purpose, and a permit or license is needed to catch aquarium fishes.

Coastal Protection

Shoreline erosion and storm damage are serious problems affecting millions of people, especially in the tropics, where homes and businesses may be built on or near the coast. Healthy coral reefs provide a natural bar-rier, helping to protect the coast by causing waves to break offshore and dampening their energy. Offshore reefs are thought to have reduced the impacts in some areas during the catastrophic Indian Ocean tsunami in 2004. In addition, behind coral reefs there are often relatively quiet-water lagoons that host important seagrass and mangrove habitats, which act as nurseries for marine life and help to further stabilize the shoreline.

Human Health

When it comes to drug discovery, coral reefs are the world's new rainfor-ests. The potential to find compounds with medical application is high because of the great diversity of species present and their common use of chemicals for defense. There are already numerous compounds on the market or in development that are derived from reef-related organisms. For example, some forty years ago, research on the chemistry of a Carib-bean sponge led to the development of AZT, one of the most important drugs used in the battle against HIV. More recently, in clinical trials the drug eribulin, derived from a marine sponge, has shown promising re-sults in extending the lives of breast cancer patients. From other sponges, researchers have also developed Discodermolide, which is in clinical tri-als as an anticancer drug, and an anti-inflammatory known as Topsentin. Disease-fighting compounds have also been found in and derived from sea squirts, gorgonians, and small encrusting creatures called bryozoans. And as noted earlier, the reef's cone snails are proving to be a medical bonanza. In addition, the growing recognition of the importance and abundance of microbes that live in association with coral reef organisms is opening new avenues of promising research. Already, a substance derived from bacteria that are found in sponges is being tested as an agent against malaria.

For years, coral skeletons have been used in bone grafting, and in Aus-tralia, scientists are working to create a sunscreen based on the natural compounds in corals that block ultraviolet light. One might also argue that there is a significant positive impact on human health from the reduction of stress obtained by a day of snorkeling or relaxing on a coral sand beach. To prevent overharvesting, responsible companies looking to use or make drugs from coral reef organisms make a requirement of their research the ability to synthesize the relevant compounds in a laboratory or to produce them through aquaculture.

Not all products derived from the coral reef, however, are necessarily good or better than others. Makers of expensive coral calcium or coral water products have advertised them as near wonder drugs, with effectiveness against cancer, Alzheimer's, and more. Experts have found no supporting scientific evidence for such statements and the Federal Trade Commission and the Food and Drug Administration have charged some manufacturers for unsubstantiated and false claims.

Materials for Construction

In small island nations, coral reef limestone and sand may provide the only material available for construction. Around the world, sand derived from coral reefs is also used in beach renourishment projects, and ancient fossil reefs can supply building materials as well. In the Florida Keys, many of the area's older homes and hotels are built out of blocks cut from a local limestone quarry that is now preserved as a geological park. Visible within the walls of these structures are the skeletons of corals that were once part of a thriving reef some 125,000 years ago, when sea level was about 8 to 10 meters (20 to 30 feet) higher than today.

Ocean Diaries

The banding in a coral's limestone skeleton is an annual diary of sorts for the oceans and climate. Scientists are now able to use structural and geochemical analyses of a coral's skeleton to determine its age and to obtain information about the seasonal as well as yearly conditions in which it grew. A large, undisturbed coral head can provide information about hundreds of years of growth, and in places, it may precede recordings by instrumentation. In Florida, cores taken through corals have provided a record of cold-front passage and a history of hurricanes over about half a century. In Australia, coral cores have been used to study the history of summer monsoon rainfalls and coastal runoff. In the Galapagos, a coral core spanning 367 years of growth between 1586 and 1953 was used to determine that El Niños typically occur about every three to seven years, and that the severe 1982–1983 El Niño that caused widespread coral mortality in the region was the most destructive in four hundred years. Fossil or uplifted coral reefs have also been used to reconstruct changes in sea level or to document earthquake or volcanic activity. Researchers continue to study corals as proxies for climate change and use their skeletons as another means to better understand our planet's past.

Coral reefs cover less than one quarter of 1 percent of the ocean, yet they harbor one of the greatest menageries of life on the planet and are an immensely valuable asset to society. For the creatures of the sea, coral reefs provide critical housing, protection, food, and a place to meet a potential mate and produce a new generation. The same could be said for humans. Coral reefs provide protection for the shoreline and in some places, housing materials. They supply millions of people with essential food, livelihoods, effective medicines, and a recreational or vacation setting and mood that are conducive to mating and reproduction. Put simply, coral reefs are places of immense wonder and enormous value; unfortunately they are also vulnerable communities whose future is now at great risk.

5 Armed and Dangerous

From the coral reef to the deep sea, there are some creatures that are particularly well equipped for hunting, with multiple appendages and predatory intentions—they are both armed and dangerous. With gripping, suckered, or stinging arms and an arsenal of weaponry that includes sharp parrot-like beaks, paralyzing venom, and dual stomachs that enable out-of-body digestion, these animals pose a serious threat to their prey. Along with an effective offense, they are also experts at defense. To ward off would-be predators they employ camouflage in 3-D, toxic or glowing slime, and a smokescreen of ink that can distract an attacker or leave it in a chemical haze, dazed and confused. The residents of the sea that are armed and dangerous have evolved abilities that are both figuratively and literally stunning.

PUMPED UP AND HUNGRY

Sea stars are beautiful, colorful creatures, well known for their radial symmetry and numerous arms. They may seem like docile creatures, lying peacefully wedged within the rocky crevices of a tide pool or sitting quietly on the mud, deep in the abyss. But their appearance often belies a dark side, as they can be mobile and rapacious killers. Sea stars also possess superhero-like powers, including the ability to scale vertical walls and heal themselves almost miraculously.

At least two thousand species of sea stars populate the world's oceans, from the shallow intertidal zone down to depths of some 7,000 meters (about 23,000 feet). They come in assorted colors and forms, from striped to polka-dot, spiny or smooth, and slender or plump. Sea stars often have five arms radiating from a central disk, but they can actually have as many as fifty limbs. Some sea stars can grow additional arms as they get older, or lose one and then add another. They have an amazing ability to regenerate. If an arm is injured or becomes trapped in the jaws of a predator, it can be self-amputated and then regrown. In fact, their power of regeneration is so great that some species can reproduce by simply splitting themselves in half and then regrowing the rest of their bodies. Like the crawling, severed hand of a bad horror film, some sea stars can also regrow their bodies from a detached limb, if it has broken off with a section of the central disk attached. The unusual body design of the sea stars is partly responsible for this incredible talent for self-repair and replication.

The skin of many sea stars is covered with tiny pincers that repel boarders and keep their bodies free of algal growth. Lying below is a skeleton made up of calcium carbonate plates with connective tissue that has special shape-shifting properties (like their relatives the sea cucumbers). When a sea star moves, its connective tissue becomes soft and pliable, allowing for flexibility and locomotion. But if a sea star needs to wedge itself within a crack or anchor itself to the seafloor, its connective tissue turns stiff, enabling it to efficiently hold on without the use of energy-sapping muscle contractions. The sea star is a brainless creature, but its lack of wit is made up for by a sophisticated internal waterworks system. Water-filled canals run throughout its body and extend down each arm. The sea star uses water pressure within these canals to extend numerous stringy "tube feet" out through grooves on the underside of each arm. An individual sea star may have thousands of tube feet and each is tipped with a tiny suction cup and can excrete sticky mucus. Their tube feet are used in a grip-and-pull-type movement, allowing the sea stars to become the Spider Men of the undersea world—able to easily crawl up steep rock walls or perch at precarious angles.

A sea star periodically refills its internal hydrodynamic system with seawater through a small pore lined with grooves and small hairs that filter out particles and prevent clogging. Scientists recently discovered that in the intertidal zone of the Pacific Northwest there is at least one sea star that likes to pump itself up during high tide. By taking in lots of water it can stay cool and hydrated even when exposed at low water. For a person, it would be like drinking numerous liters or a couple of gallons of water in preparation for an extended bask in the midday sun. Having slightly modified seawater for blood is also advantageous if injured—the sea star has no need to staunch the flow or get a transfusion. Also handy are the duplicate internal organs in their arms. While their digestive organs are located primarily within the central disk, they too branch slightly into each arm. At the underside of the central section is the sea star's mouth and atop is its anus. Next time you happen upon a lovely sea star, looking down upon its brightly colored surface, you will actually be staring at its butt.

Most sea stars are avid hunters, though some are especially picky about their prey, preferring a specific diet of clams, sea cucumbers, or crabs. Other sea stars are undiscerning diners that will eat just about anything they can catch and consume. Sea stars are especially efficient consumers of shellfish, which they eat with great style, but poor table manners—exuding of your stomach while dining is hardly good etiquette. Once a sea star has captured a clam or abalone, it wraps its arms around it and then uses its tube feet to pry apart the shell. When a slight gap opens up, the sea star everts one of its two stomachs out of its body and into its prey's shell. There, the sea star's stomach secretes digestive enzymes that transform the soft-bodied organism into a soupy liquid that can be readily slurped up. It may take hours for a sea star to completely liquefy and devour a shellfish, but when the dining is done, it simply pulls its stomach back into its body. Gives a whole new meaning to the concept of dining out—outside of your body that is.

Along with a powerful sense of touch, sea stars have the ability to detect undersea "smells" or chemicals, and they have primitive eye-spots on the end of each arm that are sensitive to light. Using these senses, some species dig for clams, while others will literally chase down their prey; and with a multitude of tube feet, they can be relatively swift of foot. An abalone will actually dart away if a sea star approaches, swiveling its body back and forth in an attempt to shake off its attacker. Even sea stars are wary of sea stars, especially those species with a penchant for cannibalism.

One of the largest and most voracious of the sea stars is the sunflower star, commonly found in the Pacific Northwest. It may be purple, reddish brown, or yellow and can reach 90 centimeters (3 feet) in diameter with up to twenty-four arms, each of which may be 40 centimeters (16 inches) long. This sea star is also a heavy sucker, and can weigh up to 5 kilograms (11 pounds). It can live within the intertidal zone or in depths of more than 100 meters (328 feet). The sunflower star has some fifteen thousand tube feet and especially soft connective tissue that provides flexibility and ease of movement over the bottom, making it a world-class sprinter among its peers. It is a limp but relatively fast creature that feeds on just about whatever it wants. It can exude its stomach to slurp up shellfish or simply drape its large body over softer prey and consume them whole. If the sunflower star injures an arm, it triggers the release of a chemical that can cause it to drop the wounded limb. This chemical also spreads into the surrounding water, where it acts like a warning signal or alarm to the other sunflower stars in the neighborhood.

In the Southern Ocean, there is a challenger to the sunflower star for the title of largest, strangest, and most efficient killer of its kind. It is a huge, clawed, multiarmed monster that feeds on relatively fast-moving prey—krill. The Antarctic sea star, *Labidiaster annulatus*, can reach up to 0.6 meters (2 feet) across, have over fifty tentacle-like arms, and as expert Christopher Mah puts it, "is covered with claw-shaped pincers, like tiny bear traps." This creature fascinates Mah, who notes that their feeding strategy is quite unusual for a sea star. A *Labidiaster* finds something to perch on, such as a tall sponge, and then extends its long, waving, weaponized limbs into the water. When a small, unsuspecting krill or other creature gets a little too close, the sea star attacks, using its clawed skin and arms in a deadly grasp. Death comes quickly as prey are moved to its mouth and promptly devoured. Photographs reveal that these are not rare, deep-sea monsters, but may occur in abundance, with hundreds perched at the seafloor fishing for krill.

Sea stars are actually quite acrobatic and versatile creatures. Some species will fence and wrestle among themselves for dominance, while others can somersault back into place if overturned. The cushion star, also known as the slime star, secretes copious amounts of mucus to repel attackers, while other species use slime to ensnare organic material as it drifts by. There are scavenging sea stars as well as some that prefer to sieve through the sand and mud for their food. Few organisms seem to prey on sea stars, although the Alaskan king crab appears to find them ripe for a meal, as do giant triton snails, several fishes, and a few seabirds.

For most of the many-armed stars of the seafloor, sex is a big release. Millions of eggs and sperm are discharged into the water, where they mix, and fertilization occurs. Young sea stars begin life as odd, bilaterally symmetrical drifters within the oceans' currents. They later settle to the seafloor and undergo a metamorphosis into their more familiar radial form. A few sea stars brood their eggs and produce hatchlings resembling miniature adults. Then again, for some sea stars it is even simpler: just cut yourself in half, regenerate your other side, and from one, you get two—clones.

STINGING GOBS OF GOO

They have no brain, blood, heart, or spine, and are composed principally of gelatinous goo that is 95 percent seawater. They are also, for the most part, ineffective swimmers. Yet jellyfish still manage to pose a serious threat to many marine organisms, and they are universally feared by humans. It is not their guile, wit, or speed that we dread, but the thousands of stinging cells that line their many long and nearly invisible tentacles. These simple organisms have been around for hundreds of millions of years and have evolved into a diverse and fascinating array of somewhat scary but beautiful creatures.

The classic jellyfish has two layers of thin tissue sandwiching a glob of watery goo, which together form a transparent dome-shaped bell or umbrella. The size of their gelatinous bodies varies from that of the tiny thimble jellyfish at less than a few centimeters (an inch) across, to the monstrous bell of the gigantic lion's mane, which can be some 2.4 meters (8 feet) wide. A jellyfish is radially symmetrical, and typically, at the base of its bell has a central mouth that opens into the stomach for digestion. Hanging from the edges of the bell are tentacles that can be just a short fringe or trail far behind for some 30 meters (100 feet). Their tentacles are lined with venom-filled nematocysts, the same type of stinging cells as their cousins, the corals. Each nematocyst contains a syringe-like needle attached to a coiled thread, which fires and injects venom upon contact with prey or associated chemicals. A jellyfish's tentacles may house thousands to millions of nematocysts, the potency of which varies with species. Jellyfish wield these tiny, toxin-laced spears to paralyze prey and make dining more manageable while drifting in the open sea. Most jellyfish also have frilly feeding arms that hang down from the middle of their bells and are used to bring food to their mouths. They are all carnivorous and feed principally on zooplankton, small fishes, and other jellies. Compared to more active, complex creatures, they require less food to fuel growth and sustain their simple bodies built of goo. The jellyfish's transparent, gelatinous body is not only low-maintenance, but also provides a dual-purpose invisibility cloak, hiding it from both prey and predators. Most jellyfish simply drift through life, waiting to literally run into their food.

Some two hundred species of jellyfish are recognized today, but scientists believe that there are many more yet to be discovered. They are found in all of the world's oceans, from the surface down to depths of over 7,000 meters (23,000 feet), from the cold of the poles to the heat of the tropics. There are many variations on the simple jellyfish theme, ranging from delicately elegant forms to those that are oddly outfitted or downright terrifying. The common moon jelly has a pinkish, translucent, dome-shaped bell and tends to drift gracefully near the surface. It is actually more of a slimeball than a serious stinger. The moon jelly is covered with sticky mucus that it uses to trap plankton, which slide down its bell and into its not-so-friendly tentacles and feeding arms. It has a relatively weak sting. Another relatively common jellyfish is the sea nettle, with long spaghetti-like reddish tentacles, ruffled white feeding arms, and a yellowish to transparent bell (plate 12). It packs a moderate to relatively strong sting. One jellyfish that is often found in shallow, sunlit waters is a bit of a black sheep among its clan; it likes to hang around, upside down and on the bottom. It appears more like a frilly flower than an armed glob of goo. Its disguise is revealed only by the weak pulsations of its saucer-shaped bell, which it uses like a suction cup to sit on, upside down on the seafloor. The jelly's underside, which becomes its topside, is covered with small tentacles and numerous frilly feeding arms; embedded within are symbiotic algae, which give it a greenish hue. The upside-down jelly farms these algae and fosters their photosynthetic growth to supplement the food it catches on its tentacles. It is also a mouthy creature, hosting not just one mouth at its center, but a whole host of minimouths on its feeding arms. And when threatened, this jellyfish launches stinging grenades into the overlying water by releasing some of its nematocysts.

Some of the most ornate and bizarre jellyfish live in the oceans' mid- and deep-water realms. With translucent orbs surrounding crimson-colored stomachs, these creatures often resemble delicate Christmas ornaments. A fringe of tentacles dangling off their bells adds to the decoration. The red stomachs of midwater jellies are thought to have evolved as a means to hide bioluminescent prey that have been consumed. In the oceans' dark depths, having a stomach with glowing contents would be like having a brightly lit, flashing sign saying, "Eat me, eat me!" Some species, however, use bioluminescence to lure in prey and have their own year-round holiday lights that they can turn on and off: glowing spots that flash or pulsate on or around their bells. Other midwater jellyfish have glowing tentacles that, when under attack, can be jettisoned as a defensive countermeasure.

Recent expeditions in the Arctic have found a surprising abundance of jellyfish living under the ice, including the giant lion's mane, a new small, blue species, and one with a golden-red stomach that holds its tentacles up to fish rather than trailing them behind.

Box jellies, sometimes called cubomedusa, include the marine stinger, sea wasp, and irukandji. Among the jellyfish, they are the most voracious predators and some of the deadliest animals on the planet. In general, box jellies can be found throughout the tropics and subtropics, but sea wasps typically occur just in the nearshore waters of some parts of Australia and Southeast Asia. In the last century, some eighty people have died in Australia due to encounters with sea wasps. Their venom is so potent that it can kill a human within four minutes if untreated.

Along with their dangerously potent sting, box jellies have some rather unusual attributes, for a jellyfish. Most jellies have a simple nerve net and cells along the edges of their bells that are sensitive to light and gravity, enabling them to stay oriented and upright. They also typically have receptors sensitive to touch on their tentacles and around their mouths, along with specialized cells on their bodies for smell and taste. Each side of the cube-shaped box jelly, however, is additionally outfitted with a cluster of six eyes; that means a deadly stinging glob of goo with twenty-four eyes. In each cluster of eyes there is one pair that is oddly sophisticated and camera-like, with a lens, retina, iris, and cornea. Scientists believe that these eyes provide the box jelly with a sort of blurry vision that allows it to avoid relatively large, stationary objects and to see prey. How such vision is processed remains a mystery, since the box jelly has no brain!

The box jelly's tentacles are ringed with concentrated nematocysts and characteristically hang in bunches from the lower four corners of its square-shaped bell. A box jelly can grow to be basketball-size, with up to sixty tentacles that are some 4.6 meters (15 feet) long, containing millions of venom-filled nematocysts. One large sea wasp can hold enough venom to kill sixty people. These not-to-be-messed-with creatures are also unusually and frighteningly strong swimmers. At best, most jellyfish are able to pulse their way slowly upward. The box jellies, however, are relatively agile and jet-propelled; they can swim in bursts of speed up to 1.5 meters per second (5 feet per second). Luckily, the box jellies' meals of choice are small crustaceans, fishes, and worms. Unfortunately, for those who are unlucky enough to run into a box jelly, the encounter can be excruciatingly painful and potentially fatal. Secure clothing can protect a person from their sting, and now, at least, an antivenom is available.

When it comes to sex, there is at least one species of box jelly that engages in a little romance and has intimate relations. The males of the small box jelly *Carybdea sivickisi* have been observed to court females in a ritual likened to a "wedding dance," though it seems more like the caveman drag-by-the-hair version of courtship. The male box jelly swims around a female, grabs her tentacles, and begins dragging her about. He then pulls her close and uses one of his tentacles to pass a package of sperm to her. She ingests his seed, uses it to fertilize her eggs, and in a few days produces a strand of thousands of embryos. It is a *touching* and unusual means of reproduction for the gelatinous crowd.

Most jellyfish live less than one year and in that time they reproduce through a somewhat less intimate, though complex, two-act life cycle. For the majority of male jellyfish, sex involves the release of sperm into the water, which then fertilizes eggs that have been discharged by a female. Alternatively, a female jellyfish may capture the released sperm and use it to fertilize eggs which she is carrying. Fertilized jellyfish eggs then develop into larvae that temporarily become drifters; they will eventually settle and attach to the seafloor. There they morph into a flower-like polyp structure that buds asexually—like a mother ship releasing tiny flying saucers from a stack, one after another. These are the baby jellies that grow into the familiar floating bell-shaped forms. Interestingly, researchers now believe that the budding phase of the jellyfish can be delayed or remain dormant until favorable conditions for the young develop, such as warm temperatures or an adequate supply of plankton for food. Fields of tiny, nearly invisible polyps, potential jellyfish factories, may sit attached to the seafloor just waiting for the right time to release a brood of babies. This may partially explain why jellyfish are often found in swarms and how their populations seem to bloom periodically.

Jellyfish, even the most potent of the bunch, are not immune from predation. Sea turtles the world over are one of their main predators. Even the nefarious twenty-four-eyed, fast-swimming, power-stinging box jellies are sumptuous meals for the sea turtles. Other creatures that consider the jellies delicious gummy treats are the ocean sunfish, tuna, butterfish, and the spiny dogfish.

There is also a jellyfish imposter lurking in the sea, drifting at the surface. It is a balloon-toting, blue-tentacled creature that packs a strong sting and resembles a jelly. But the Portuguese man-of-war is only masquerading. It is not a true jellyfish, but rather a siphonophore, which is made up of a colony of individuals or polyps—a floating commune of sorts. Other types of siphonophores also drift throughout the worlds' oceans, and their colonies can reach extraordinary lengths, some 40 meters (130 feet) long. Different types of polyps make up a siphonophore. In the Portuguese man-of-war, one polyp forms its characteristic purplish-blue gas-filled bladder, which floats at the surface and can reach up to about 0.3 meters (1 foot) in length. This balloon-like member of the community keeps the rest of the group afloat. Underlying the bladder-polyp are the food processors or feeding polyps, a set of community mouths and bag-like stomachs. Another set of individuals constitute the man-of-war's tentacles; they are the food gatherers or hunters of the group and can reach up to some 12 meters (40 feet) in length. The tentacles are beaded with nematocysts whose venom is nearly as potent as that of a cobra. Unlike a true jellyfish, the Portuguese man-of-war has muscle along the length of each tentacle that can contract and literally reel in prey, thereby delivering it directly to the feeding polyps. When prey such as small fishes and crustaceans are caught and digested, nourishment is provided throughout the community. In addition to the individuals that make up the Portuguese man-of-war's gassy float, hungry mouths, and writhing tentacles, there are polyps that form its reproductive structures. There may be just a few or up to a thousand organisms making up the commune that is a siphonophore, such as the Portuguese man-of-war.

Sea turtles and a few fishes are able to consume the Portuguese man-of-war, and as described earlier, at least one sea slug not only eats man-of-war tentacles, but also uses their unfired nematocysts in defense. Pacific sand crabs are also known to feed on the man-of-war. When one is washed ashore, a crab will drag it onto the beach and start burying it, presumably in preparation for a feast. If one clawed crustacean cannot do the job, mob mentality takes over and more crabs come to help.

The Portuguese man-of–war sometimes picks up a nibbling, ocean-going hitchhiker, the small, striped man-of-war fish. As a juvenile, it is a stowaway among the man-of-war's stinging blue tentacles, where it gets protection from potential predators and a steady stream of snacks. It nibbles on the siphonophore's prey as well as its tentacles and reproductive organs. Hitchhikers also like to catch a ride with jellyfish. Small creatures such as crabs, amphipods, or barnacles may ride atop a jelly's bell or, like the Pacific butterfish, swim among its stingers. It is unclear how fishes that dance amid the stinging tentacles of the world's jellyfish become relatively immune to their venom (slime perhaps?), but it certainly gives them an advantage in the eat-or-be-eaten world undersea.

LIVE FAST AND DIE YOUNG

Among the organisms of the oceans that are armed and dangerous, there is one elite group of jetsetters that grow fast and are blessed with brains, looks, and a talent for disguise. Though most don't live long, they can react with lightning-quick speed and deploy amazing defensive countermeasures. They are the stars of film and literature, playing oversized and aggressive monsters with a killer appetite. Most of the oceans' cephalopods, however, are not out for a fight; when trouble calls, the majority of squids, cuttlefishes, and octopuses prefer to hide or run away. They are also much stranger in fact than in fiction, with truly astonishing capabilities and many bizarre characteristics. Their diverse ranks have come a long way from their distant and rather simple cousin, the clam.

More than eight hundred species of cephalopods reside within the world's oceans and include squids, octopuses, cuttlefishes, the nautilus, and a unique deep-sea crossover—the vampire squid. They are all mollusks, with a heredity distantly connected to shellfish such as clams, oysters, and snails. Oddly, what was once a foot evolved to become mostly a head along with a bevy of arms, and all but the nautilus lost their protective outer shells.

Cephalopods grow fast and most reach maturity within two years. To fuel their rapid growth, most squids, octopuses, and cuttlefishes are voracious and well-equipped hunters, feeding on live prey, such as crustaceans, fishes, shellfish, and other cephalopods. They have eight muscular, suckered arms, which they use to grab and immobilize their victims. Squids and cuttlefishes also have two longer, elastic tentacles that may be tipped with hooked clubs and can be deployed in an instant to lash out at potential victims, aka food. Suckers on a squid's tentacles may be teethed for better gripping. Because cephalopods have relatively small mouths, these hunters must chop, drill, or soften up their catch before swallowing. To do this, they often use their hard, parrot-like beaks, which can tear apart flesh and crush bone. Within their mouths, they also have rasping tongues lined with small teeth, which some octopuses use to drill through their prey's outer shell or exoskeleton. And cephalopods have acidic, venomous spit. Their salivary glands produce digestive enzymes and toxins, the latter of which vary in potency with species. The saliva of the small blue-ringed octopus contains one of the deadliest venoms on the planet. These potential "death spitters," however, seem to be rather timid creatures and use their potent saliva mainly for feeding. A cephalopod's spit is typically used to paralyze prey. When feasting on crabs, an octopus wields its spew with purpose. It drills a small hole into a crab's carapace and then injects saliva to destroy its prey's attachment muscles and begin the digestion process. And being picky, fastidious eaters, octopuses make the most of each meal. Piles of disarticulated shells or crustacean carapaces just outside an octopus's lair are often picked clean of meat.

Most cephalopods use their excellent vision to hunt down their prey, but they are also able to feel out their victims or seek them through "smell" or chemical cues in the water. Some are also able to sense vibrations produced by an organism's movements. Once prey is located, cephalopods are well outfitted for the chase. In the open ocean, a squid can outswim and outmaneuver many other animals. Its streamlined, torpedo-shaped body includes a jet-propulsion system that uses muscular contractions to force water into its mantle or bag-like body and out through a maneuverable pipe-like siphon, enabling the squid to become an agile, speedy jetfighter or stealthy hovercraft. Fins along its sides or rear are used for steering and stability, and sometimes for swimming. And when they go from hunter to huntee, some species of squid can even rocket out of the water and glide for up to 50 meters (164 feet)—though Hollywood has yet to produce *Attack of the Killer Flying Squid*. Though probably not as fast or as agile as squids, many cuttlefishes and octopuses also rely on in-body jet packs for swimming. In contrast, deep-water octopuses use fins for slow-motion sculling, and have webbing between their arms that acts like a parachute or sail. Octopuses that live on the bottom are extremely mobile creatures as well, which can crawl over just about any obstacle, can walk on their arms if need be, and are expert contortionists, able to squeeze through incredibly small openings while on the hunt or to avoid predators.

Hunting strategies vary among the cephalopods. An octopus may grab or pounce on its prey and then drag it back to its lair for stockpiling. As tactile feeders, they may poke and prod looking for food, or simply spread their arms or webbing out and eat whatever they find and capture. Other cephalopods, such as the squids, stealthily stalk their prey or choose a more direct approach and attack head-on, like a ballistic missile strike. With excellent skills at camouflage and mimicry, some of this group may also lure in unsuspecting prey. One deep-sea octopus has special suckers along the length of its arms that emit blue-green light. They may glow dimly or blink on and off, and are thought to attract potential quarry. Beneath its webbed arms, this octopus also produces mucus (seems like under the sea almost everyone does), which may ensnare small crustaceans that drift or swim into it, like a slimy net.

While on the hunt, cephalopods must always be wary of predators. Having lost the protection of an outer shell, their naked, soft bodies offer energy-rich, enticing meals. Just about every type of marine carnivore eats cephalopods, including whales, dolphins, seals, seabirds, and fishes such as the billfishes, tuna, groupers, and sharks. Even the cephalopods eat cephalopods.

To avoid the oceans' hungry masses, cephalopods have developed an extraordinary array of defenses. Roger Hanlon of the Marine Biological Laboratory in Woods Hole, Massachusetts, has been studying cephalopods for decades. He discovered that when confronted, their first line of defense is camouflage, and that they are the fastest, best-equipped animals on the planet when it comes to hiding in plain sight. Occasionally, even he has been fooled by their amazing ability to disappear into their surroundings. Cephalopods are able to quickly and precisely take on the appearance of the bottom and become very uncephalopod-like. They can match the brightness, color, pattern, and texture of the seafloor in seconds, create optical illusions, and change the shape of their bodies to mimic in 3-D. Much of their success as disguise artists lies in their sophisticated "smart" skin.

Scattered throughout a cephalopod's skin are thousands of small color organs called chromatophores, essentially little muscular, elastic sacs containing pigment. When muscular contractions stretch these pigment sacs into flat disks, a cephalopod's skin is visibly colored or patterned. Individual or groups of chromatophores can be contracted at a time, creating an astonishing assortment of hues and designs. The cephalopods' prowess as dress designers is not only impressive—it is also fast. In the blink of the eye, they may become striped, polka-dotted, or covered in psychedelic waves of color. An octopus may appear as if painted ruby red one minute and within seconds go ghostly pale or become a dull, mottled brown that replicates the underlying rock or sand bottom. The blue-ringed octopus can cause its namesake circular markings to flash brightly, and many cephalopods can create waves of color that wash over their bodies like a passing cloud. They can also produce patterns that disrupt the outlines of their bodies so that they become nearly unrecognizable. Surprisingly, their chromatophores contain only red, orange, yellow, black, or brown pigments, yet cephalopods exhibit a full rainbow of colors. The mystery of how they do this was solved when scientists discovered that they also have reflecting cells in their skin that act like mirrors or prisms. These iridophores and protein-based leucophores are used in combination with their chromatophores to create a wide variety of striking colors that include vibrant blues, purples, greens, and silver. How cephalopods respond so quickly and create so many patterns is, however, about more than just the color of their skin.

An intricate nervous system runs throughout a cephalopod's epidermis, connecting its colored pigment organs and reflector cells to its relatively large brain and complex eyes. They are, in fact, the brainiest of all invertebrates, having the largest of the group along with especially well developed eyes. Hanlon's research team has discovered that cephalopods use their exceptional vision as their primary means of detecting the brightness and patterns within their surroundings, which they then quickly replicate for camouflage. But, ironically, Hanlon's team also found that most, if not all, cephalopods are color-blind. How then do they perfectly match the color of their surroundings? He suspects that the cephalopods' skin has some sort of color-sensing capability, but what it is and how it works remain unknown.

As true masters of disguise, octopuses and cuttlefishes can also change the texture of their skin, creating bumps, ridges, or algae-like frills; this too appears to be a vision-based skill. Posing perfectly still or moving in an uncephalopod-like manner, they can imitate their surroundings in 3-D, becoming part of a rock, hiding among seaweed, or "impersonating" another organism. Octopuses have been seen to change their bodies to look and move like a flounder, a sea snake, or a drifting tumbleweed of algae. By hovering motionless and pointing their arms upward, a squid can become nearly unrecognizable within a stand of algae. Within the animal kingdom, the cephalopods' rapid camouflage capabilities are simply the uncontested best. Even the iconic chameleons cannot match their speed or capabilities as quick-change artists.

Some cephalopods also have an illuminating means of camouflage. Using photophores or light organs on their undersides, squids can produce light that matches the radiance downwelling from above. This counter-illumination renders them invisible to predators looking up from below.

If camouflage fails, the cephalopods have several other tricks up their many suckered arms. They may attempt to bluff or startle an attacker with a display of flashing color, a rapidly changing pattern, or the release of ink. Many squids, octopuses, and cuttlefishes can eject ink as a smokescreen to facilitate a quick escape or as more of a decoy when released in combination with mucus to create a cephalopod-like shape. In the oceans' darker depths, a cephalopod's decoy or smokescreen may be bioluminescent. Some scientists speculate that this may provide an additional advantage: if an attacker gets smeared with glowing slime, it may go from stealthy predator to obvious prey, as the hunter becomes the hunted. Cephalopod ink may also irritate the eyes of other animals and chemically cause an attacker to become confused and disoriented. If camouflage or counter-measures don't do the trick, cephalopods may exhibit erratic or unpredictable behavior in hopes of confusing predators. And if none of these tactics work and a hunter successfully nabs an arm or two, cephalopods are able to regenerate lost appendages. Some may even be able to lose a limb on command, if necessary.

The cephalopods' exceptional shape-shifting and color-changing skills also provide an important means of communication. A certain posture or display may be used to attract mates or to ward off competing suitors. Color changes or positioning can signal aggression or lust, or sometimes both at the very same time. A male squid courting a female may blanch seductively on one side, while his other half darkens as a threat to watching rivals—literally exhibiting two moods at once.

In cephalopods the sexes are separate and they are all promiscuous. Fidelity has little place in their typically short lives; most live only one to two years. They are mainly seasonal breeders and often kick the proverbial bucket after reproduction. Some mate just once in a lifetime, while others get it on repeatedly during a single season, and a few species are able to reproduce throughout their ever-so brief lives.

Octopuses, for the most part, are solitary creatures and sex happens with little romance or foreplay. When a male octopus is ready and feels the need, he comes out into the open and may flash with color, display enlarged suckers or spread the web between his arms. But that is the extent of the preliminaries, as he may then leap on a nearby female or reach around to her from a distance with his specialized baby-maker arm. Male cephalopods are equipped with a characteristic arm that is designed especially for sex and used to transfer a package of sperm to females (plate 13). Female octopuses are not always willing partners, though, and are known to often reject suitors. On rare occasions, multiple male octopuses may compete for a single female and more than one may try to mate with her at the same time. In that situation, eight arms might just come in handy. Because octopuses can be cannibalistic, small males must be wary when coming out into the open to mate—for them attraction truly can be fatal.

After mating, a female octopus typically lays her eggs in a den and stands guard. She gently jets water over her brood, keeping them clean and aerated, and forgoes food to ensure the safety and development of her eggs. Once they hatch, her parental duties are done; she is spent, and dies. Male octopuses tend to die at about the same age as the females. Theirs is the ultimate sacrifice for their young. But for at least one small, cold-water octopus, the female's parental duties are not a fast finish to the end. Scientist James Wood at the Aquarium of the Pacific discovered that the female of the octopus *Bathypolypus arcticus* protects and broods her eggs for over a year, all the while wasting away as she metabolizes her own body to provide energy so that she can care for her offspring. This species, which lives in very cold water, is also unusual in that it can apparently live a relatively long life for a cephalopod, perhaps up to four, possibly even six years.

There are some octopuses that live in the open ocean and do not have dens at the seafloor in which to raise their young. In these species, females may carry their eggs cradled in their arms until their newborns emerge. The blanket octopus is one such species and spends its entire life drifting throughout the open ocean. It is an odd creature, even for an octopus, and has been likened to the caped crusader of the seas. It is typically dark pink, blue, or purple, with extended maroon webbing between its arms that trails behind it like a blanket or cape. And in this octopus, the sexes really are different—extremely different. The females may grow to be up to about 2 meters (6.5 feet) in length, whereas even mature males are just 2 to 3 centimeters (1 inch) long. Biologist Tom Tregenza at the University of Leeds likens it to a human-sized female mating with a male the size of a walnut. When it comes to pleasing their mates, however, the diminutive males of this species loom large and are especially self-sacrificing. During mating, the male blanket octopus not only passes a package of sperm to his mate using his specialized arm, he actually gives her part of the arm—it is detachable. Females have been found with the arms of numerous males in their mantles or body cavities. They are collectors of sperm and mates, at least of their body parts. After the male blanket octopus self-amputates part of his arm during sex, he dies. The males are the martyrs of their species, while the females sleep—that is, swim—around.

The diminutive males and immature females of the blanket octopus have also evolved a stunning way to deter predators. If a Portuguese man-of-war drifts by, they reach out and tear off a few tentacles, which they then carry clutched within their suckers at the ready for use in warding off attackers. Somehow, they too are immune to the man-of-war's sting.

Reproduction in squids and cuttlefishes is distinctly different from that of the solitary, standoffish octopuses. During mating, male squids and cuttlefishes court and often guard their partners, at least temporarily. Also in contrast to the octopuses, female squids and cuttlefishes do not tend to their eggs once laid, and fertilization takes place externally. Most cuttlefishes breed in pairs or small groups over an extended area, but every winter in the rocky reefs of southern Australia, thousands of giant cuttlefish congregate in an unusual mass orgy of bottom-dwelling cephalopods. Hanlon's team has spent hundreds of hours documenting this gathering and the associated reproductive behavior of the giant cuttlefish, which is unique to Australian waters. It has a flattened body that reaches up to about 50 centimeters (1.5 feet) in length, but with the extension of its eight thick arms and two feeding tentacles, it can appear even longer. Its eyes are big, and a ruffled fin surrounds its mantle or body. As with most of its kin, the giant cuttlefish also has exceptional color-changing and camouflage abilities.

Around late April, giant cuttlefish begin arriving at their breeding site in southern Australia, and by the peak of the gathering there may be over forty thousand individuals in the area. Among the gatherers there are more males than females, leading to intense competition for breeding partners. Most male rivalries are settled through aggressive displays or posturing, which signals their gender and readiness to fight. On rare occasions, the competition can get physical. Both sexes mate with multiple partners, their rendezvous brief. The females, however, have the upper arm, deciding which males get rejected and which get lucky. They may reveal their choice by arching their bodies in a rejection or spreading their blanched arms toward the males as an acceptance. At their first attempt, many male giant cuttlefish are sent packing. If not, a male spreads his arms and grasps the sides of a female's head and she reciprocates, opening her arms and overlapping his in a sexual, but maybe not loving, embrace. The male jets water over the female's private parts in an attempt to dislodge any sperm deposited by previous mates. Then, using his specialized arm, he makes his move, transferring a package of sperm to the female.

After copulation, the male giant cuttlefish alertly stands guard as the female readies to lay an egg, searching for the underside of a rock to deposit one of her brood. Other males may challenge the consort male or try to sneak in to plant their own seed. Hanlon's team identified three tactics that these "sneaker" males use to get access to a female. They may simply hang around until a guarding male is distracted by a challenger or they might hide within a potential egg-laying site, lying in wait for an unsuspecting female to arrive. An even sneakier tactic is used by small males that hide their long baby-maker arms and adopt the coloration and posture of females. By appearing as members of the opposite sex, the small males lull the guarding males into a sense of security to get access to their mates, while also avoiding the aggression of other males. Some of these female "impersonators," however, are a little too convincing for their own good, and males may begin to guard them or even try to mate with them.

During the annual orgy of the giant cuttlefish, females mate with multiple males and lay eggs at numerous locations. Each day in fact a female may mate seventeen times with from two to eight different males. From her eggs, miniature cuttlefish will hatch some three to five months later and have varying paternities. This reproductive strategy gives her offspring greater genetic variety, which increases the chance that some will survive. For the giant cuttlefish, the winter gathering will not be a repeat affair, because after just one breeding season they too take their last breath, swim their last swim—and die.

Less detail is known about the reproductive intimacies of many squid species, as it is more difficult to observe their midwater encounters. Since many squids live in schools during much of their lives, reproduction is often a group affair. Males compete for females, and both sexes appear to be promiscuous, taking on multiple partners throughout a breeding season. Squids may exhibit specific courtship rituals such as paired synchronized swimming or a flirtatious game of short-distance flee and pursuit. Color changes, patterning, and arm positioning are used to signal desire, rejection, or aggression. Small male squids also exhibit "sneaker" tactics to get access to females guarded by larger males. Squid sex seems to be exceptionally brief, and it may only take seconds for a male to transfer his sperm to a female (plate 14). They produce gelatinous egg masses that may lie aggregated at the seabed or attach to a substrate or float within the water. Clusters of jelly-encased eggs can appear mop-like at the bottom or resemble large, white, elongated capsules. Depending on the species, a single female squid may produce well over a hundred thousand eggs. Incubation time varies between species and also depends on size and water temperature. Squids do not tend to their eggs, and hatchlings from a single brood may have multiple paternities as in the cuttlefishes. In some cases, there may be a large-scale mass mortality after breeding, resulting in the parents' corpses lying among their eggs at the seafloor, the previous generation amid the promise of the next. And cephalopods become hunters very early on; within just a day or so of hatching, baby squids and other cephalopods are already capable of capturing prey, albeit small to start.

As a group, the cephalopods are diverse and include some pretty strange characters, many of which live in deep, difficult-to-access marine environments. As Wood puts it, "They are not rare creatures, just rarely seen and seldom studied." Of course the most infamous example of this is the giant squid. Even though it is one of the largest animals on the planet, possibly reaching an overall length of some 18 meters (60 feet), to date it has been caught on film, momentarily, only once, in 2005. Like other cephalopods, giant squid are believed to grow fast, and may attain their immense size in just a few years. They are thought to occur throughout the world's oceans and inhabit depths mainly between 300 and 1,000 meters (980 and 3,280 feet). What we know about this huge and mysterious creature comes mostly from specimens caught in fishing trawls or that have washed ashore dead. For years, debate has raged over whether the giant squid is a strong, muscular, and fast predator or more of an oversized weakling. Their reputation as being a bit on the limp side stems from the fact that their arms and bodies, like those of many mid- and deep-water squids, are known to contain a solution of ammonium chloride, which helps to keep them neutrally buoyant. Many scientists, however, are rethinking the weakling theory based on the giant squid's one on-air appearance, in which it attacked and dragged a heavily laden cable. As for giant squid sex; scientists hypothesize that the females may store the males' sperm under their arms (female cephalopods often store sperm not only in their reproductive tracts, but also on their heads, arms, bodies, and around their mouths), and that while the male has a specialized arm for sperm transfer, he may also have a 1 meter (3 feet) long penis. How, where, and when they mate remains unknown. The giant squid appears to feed mainly on deep-water fishes and other cephalopods. Because of its fast growth rate and size, its predators are thought to be limited mainly to toothed whales, in particular the sperm whale. Reports of the two fighting it out, though, seem a bit far-fetched, as once caught, the giant squid is probably quickly bested by the powerfully built whale.

In the oceans' mid- and deep-water realms there are many cephalopod oddities. One such creature is the cock-eyed squid, resembling a giant strawberry with bizarre, asymmetrical eyes. Its richly red body is covered with tiny rows of photophores (light organs), and one of its eyes is small, bluish, and sunken within its head, while the other is larger, yellowish, and sticks out. Another strange-eyed midwater squid has one large eye looking up and a smaller eye directed downward. It can simultaneously be on the lookout for predators above as well as below. Now there is a morphology that would come in handy before crossing a busy street. During a 2005 Arctic cruise for the Census of Marine Life program, cephalopod expert Ron O'Dor observed another armed oddity, noting, "Its fins flap slowly when it swims. It looks like Dumbo." The bright pink, large-finned octopus with webbed arms made headlines as—what else—the Dumbo octopus.

In the dark depths of the temperate and tropical oceans, some 600 to 1,200 meters (1,900 to 3,900 feet) down, lives what may be the strangest of all the cephalopods. It has a velvety dark body, webbed arms, and large, ball-like eyes that glow blue or red in the light. Named *Vampyroteuthis infernalis*, or the vampire squid from hell, it is a unique "multiethnic" cephalopod with the combined characteristics of a squid, octopus, and cuttlefish. The vampire squid, however, is not a blood-thirsty predator as its name suggests, but rather a big softy that fishes using its own reels and retractable lines. Its body resembles a large, soft, brown or black football outfitted with eight suckered, webbed arms that have small fingerlike projections on their undersides. The vampire squid also has two unique tentacle-like filaments that are thin and elastic, and can be deployed like fishing lines. When reeled in, its filaments are tucked away within two small pockets between its arms. Young vampire squid use jet propulsion to get around, but as they get older, they appear to rely more heavily on the flapping of their two rounded, stubby fins. These fins develop as they mature and take the place of smaller baby fins that get reabsorbed back into their bodies. The arms of the vampire squid are especially peculiar, tipped with unusual light organs that can glow or pulse or release a thick sticky fluid containing bioluminescent particles—that is, they have arms that can fire beads of glowing slime. When the arms all fire at once, a shimmering cloud envelops the vampire squid, presumably as a defensive strategy. Researchers have also observed the vampire squid in an inexplicable and curious feat of undersea acrobatics. It swings its webbed arms up and over its head, essentially turning itself inside out, with its suckers and associated fingerlike protrusions sticking out. Witnessing this odd behavior, biologist Bruce Robison at the Monterey Bay Aquarium Research Institute describes them as going from "a soft football into a spiky pineapple with a glowing top." He believes they can see through their arms' thin webbing in this position, but why they turn themselves inside out remains a mystery. The vampire squid appears to feed mainly on small crustaceans and likes to hang out in zones of low oxygen. Staying in poorly oxygenated waters may be a means for it to avoid faster predators that have a higher metabolism and are less capable of absorbing oxygen as efficiently as they do.

Among the cephalopods, there is only one that still has the outer shell of its distant relatives. The nautilus is an oddly beautiful creature that resides within a coiled, pearly shell that the ancient Greeks considered a symbol of perfection. It has large, simple pinhole eyes and some ninety suckerless arms, and is found only in the warm waters of the Indo-Pacific down to depths of about 500 meters (1,640 feet), often along the steep slope of a reef front. Scientists believe they cannot go deeper than about 700 meters (2,200 feet) or their shells would implode. The nautilus uses its shell as both a refuge and as an unusual device for buoyancy control. If threatened, the soft-bodied nautilus can withdraw into its casing and cover the opening with a tough, leathery hood. Its home is done in camouflage with countershaded coloration, having chocolate-brown or rust-colored stripes on the top and a white underside. The shell is chambered and the nautilus resides in the last or most recently constructed section. The other chambers are filled with either fluid or gas. By adjusting the amount of gas versus fluid in its shell, the nautilus can control its buoyancy and float neutrally. It swims by retracting its body into its shell to create an awkward and weak form of jet propulsion. Unlike the others of its clan, this cephalopod's vision is poor, so it must rely more heavily on smell and touch to scavenge for food. The nautilus seems to dine preferentially on crustaceans and often migrates into shallower waters at night in search of a meal. Instead of just one specialized arm for sex, the male nautilus has several arm designates that are used in reproduction. Observations suggest that the males are indiscriminating and easily confused when it comes to the mating game. They will apparently approach and try to have sex with just about anything that appears nautilus-like. And once a male nautilus successfully identifies a female—of his own kind—he is undaunted if she lacks interest, grabbing and sometimes biting his feisty companion into a tryst that may go on for hours. For the nautilus, it is not just one season of sex and then death; they are able to mate over multiple years. This is partly because, for a cephalopod, the nautilus lives a surprisingly long life, perhaps growing as old as twenty.

WHY THEY MATTER

Sea stars, jellyfish, and cephalopods all play important roles in the oceans as predators, both in open water and at the seafloor. Cephalopods are also a particularly significant prey item, providing food for nearly every type of carnivorous creature in the sea. These organisms have also long been a part of human society and culture. For thousands of years, people have hunted and revered the octopus, using its image to adorn anything from coins to pottery and ships. European artists have historically prized the deep-brown ink of some cephalopods, and over the ages, squids and octopuses have provided an important source of sustenance. Cephalopods have also inspired our imaginations and given rise to tall tales of deep-sea monsters that have sunken ships, guarded treasure chests, and attacked unwary divers. Since 1916, cephalopods have starred in more than sixty movies, from the James Bond adventure *Octopussy* to Disney's *The Little Mermaid*, and of course, the Jules Verne classic *20,000 Leagues under the Sea* in which Captain Nemo battles the giant squid. Even the iconic King Kong has taken on a fiendish, oversized octopus. More recently, in the wildly popular *Pirates of the Caribbean* films, the horrifying kraken is the fearsome monster from the deep. Is it the cephalopod's bizarre physiology and strange looks, or its predatory ways and mysterious life that we find so fascinating, yet frightening?

Sea stars and the nautilus have had an extended history in popular culture as well, as exemplary forms illustrating the beauty and symmetry found in nature. Among collectors, the pearly, coiled shell of the nautilus is much prized, while sea stars, both real and replicas, are an ever-popular symbol of the oceans and outdoors. Sea stars adorn us as jewelry or decorate our homes on plates, napkins, and in works of art. In aquariums and at nature centers, colorful sea stars lure children to touch tanks, while the giant Pacific octopus is always a big draw for people of all ages. With the innovation of constantly circulating round tanks, even gooey jellyfish have become aquarium favorites. Their pulsating bells, trailing tentacles, and frilly feeding arms highlight these creatures' simple elegance and surprising beauty. For years, the armed and dangerous creatures of the sea have engaged our imaginations, inspired our creativity, and provided sustenance. Today, they continue to supply an important source of food and inspiration, and are also providing surprising new tools and insights in the fields of medicine and biotechnology.

Food

In Asia and along the shores of the Mediterranean, diners have traditionally feasted on squids and octopuses, and in the past few years, calamari
has risen to celebrity status, especially in the United States. Delicately bat-
tered and fried to a golden brown, squid rings have become a mainstay
on restaurant menus. Across the world, squids are not only deep-fried for
consumption, but also sautéed, stir-fried, used in salads and pasta dishes,
stuffed, grilled, and eaten raw. More than 3 million metric tons of squid
are harvested each year from the sea and in California, it has become the
state's most valuable fishery, in 2008 accounting for most of the 84 million
pounds of squid that were caught commercially in the United States. Squid
fisheries provide millions of people with an important source of protein or
a coveted snack along with jobs and economic revenue. Octopus is also a
popular menu item, though in some cultures more so than others. In Asia,
jellyfish are considered a delicacy; they may be salted and dried, prepared
as thin white pancakes, or soaked in saltwater, steamed, and seasoned. In
the United States, eating jellyfish has yet to catch on, possibly due to the
unappealing idea of chowing down on stinging gobs of goo that reportedly have little flavor and are the consistency of chewy rubber bands.

Health and Technology

In 2008, the Nobel Prize in Chemistry was awarded to scientists who discovered and developed a green fluorescent protein. Their work was lauded
for providing nothing short of a revolution in how cells are imaged,
mapped, and probed to better understand human physiology. Green fluorescent protein (GFP) is now used in a vast number of studies that range
from genetics to mapping the neural circuits in the brains of mice, and in
combating human diseases such as cancer, HIV, and Alzheimer's. It has
provided an enormous advance in the field of biomedical research that
would not have been possible if not for a creature of the sea, the organism
in which the GFP naturally occurs and was first discovered—a glowing
green jellyfish.

Sea stars have also proven valuable in the biomedical field. Early research on their larvae laid the foundation for our understanding of inflammation, and mucus from sea stars is being investigated for its use
in battling respiratory diseases, allergies, and infection. Researchers are
also trying to better understand the sea stars' and cephalopods' impres-

sive powers of regeneration and how they might apply to human ailments and injuries.

The cephalopods have, in fact, been a boon to the biomedical world. Study of the squid's giant nerve cells, likened to thin pieces of spaghetti, has greatly enhanced our understanding of nerves and the nervous system, with application to related diseases such as multiple sclerosis. The cephalopods' large brains, well-developed eyes, and sophisticated senses are also the focus of medically related investigations, along with their ability to produce venom, which may eventually aid in pain management and the treatment of allergies or cancer. The pigment within the cephalopods' chromatophores is in the melanin family, similar to the pigment responsible for color in human skin. Understanding how they synthesize these pigments may help us to better treat people with skin diseases. The success of cephalopods as research models is aided by the fact that they can often be cultivated and reared in a laboratory setting.

In biotechnology, the sophisticated smart skin of the cephalopods has intriguing potential. The nonmetal reflector cells in their skin may enable improvements in many of the products we use, from reflective road markers to fiber optics and in cosmetics. The cephalopods' ability to quickly change color and become virtually invisible in their surroundings is also of interest, especially to the military. One can imagine the military advantages of having wetsuits, clothing, or even vehicles that could quickly change color or patterns to match the surrounding environment. And if the cephalopods have color-sensing capabilities in their skin, it could lead to innovations in nanotechnology, material science, and optical devices. In the future, thanks to the cephalopods, there really might be an invisibility cloak or a car that can change color to become virtually invisible in its surroundings.

Too Much of a Good Thing

In natural numbers, the armed and dangerous creatures of the sea are, for the most part, good for the environment and rarely pose a serious threat to people. Of course, a chance encounter with a box jelly is not something to be taken lightly. But when populations of these marine organisms increase dramatically or bloom, the results, can be devastating both in the oceans and to humans.

Since about 1963, in the Indo-Pacific, population blooms of one sea star species have put the region's coral reefs and the resources they provide at risk. The crown-of-thorns sea star is a voracious killer, a purveyor of destruction that is covered with toxin-laced spines and has an appetite for coral, lots of coral. It feeds by draping its large body and seven to twenty-three arms over a coral, everting its especially large stomach, and digesting the underlying live tissue. Outbreaks of the crown-of-thorns have decimated immense areas of coral reef in the region. For example, in Guam, over some two and a half years, 90 percent of the corals in an area over 38 kilometers (27 miles) wide were killed. Infestations of the crown-of-thorns continue to periodically occur and leave behind wide swaths of dead coral. The exact cause of these outbreaks may vary by location and in some cases remains poorly understood, but factors that may contribute to the problem include overfishing of predators, removal of organisms that feed on crown-of-thorns larvae, nutrient enrichment, and climate change.

Individually, jellyfish can be a nuisance and painful when encountered, the more potent varieties dangerously so. But in a swarm, they can be a hazard to health and safety, disrupt regional fisheries, and clog intake systems for power plant cooling or desalination operations. In Japan, a ten-ton fishing trawler reportedly sank after the workers aboard tried to haul in nets that were filled with a mass of giant jellies.

Jellyfish naturally occur in swarms, but now their numbers appear to be on the rise, with blooms happening more frequently and in areas where they have previously not occurred. Waters in and around Japan have been particularly hard hit, as millions of giant nomura jellies have become un-wanted, annual visitors. They are enormous jellyfish with a blood-orange bell some 2 meters (6 feet) across, long stinging tentacles, and a large mass of feeding arms; and they can weigh hundreds of kilograms (pounds) each. Blooms of nomura clog fishing nets, decimate local fisheries, and are es-timated to cost the Japanese fishing industry over three hundred million dollars each year. Thousands of fishermen are reportedly now seeking compensation from the government. Japan is not alone in its plight. In the Gulf of Mexico, jellyfish blooms are estimated to annually cost the shrimp-ing industry some sixty million dollars. In 1999, a Philippine power plant was shut down when fifty truckloads of jellyfish clogged its intake pipes and plunged forty million people into darkness. In Iran, jellyfish have dis-rupted operations at a desalination plant, and in both the Black and Bering Seas, gelatinous marine organisms have dramatically impacted the fishing industry. And at popular tourist beaches in the Mediterranean and along the shores of Chesapeake Bay, summer jellyfish blooms are driving away beachgoers. Scientist Jeremy Jackson at the Scripps Institution of Oceanography describes the worldwide phenomenon as the "rise of slime" and warns of dire consequences as jellyfish begin to take over our coastal waters. As with sea stars, the effects of global warming, overfishing of predators, and nutrient enrichment may be making conditions more favorable for the reproduction, growth, and survival of jellyfish.

Too many cephalopods can also be a problem. Reports suggest that off the West Coast of the United States, the Humboldt or jumbo squid is becoming more abundant and aggressive and expanding its range northward, with significant impacts on the regional ecosystem and fisheries. The ocean is changing and in some cases too much of a good thing is becoming worrisome.

Sea stars, jellyfish, and cephalopods are extraordinary marine organisms with many bizarre forms and astonishing capabilities, from their high-speed camouflage skills to their hunting prowess and amazing regenerative abilities. While we have discovered much about these fascinating and at times frightening creatures, there is yet much more to learn. We don't know how many there are, where they all live, what they do day-to-day, or how their populations are changing over time. And we are just beginning to discover how much we can truly benefit and learn from them. Let us hope that the armed and dangerous continue to populate and prowl the oceans in abundance and further reveal to us their mysterious and amazing ways.

6 Cabinet of Curiosities

The diversity and abundance of life beneath the waves is bewildering, and some of the oddest creatures are among the more than sixteen thousand species of fishes that inhabit the oceans. They are wondrous animals that challenge our minds and evoke provocative questions about evolution and adaptation, or that just make us shake our heads and stare in unabashed incredulity. If not their looks, it might be their lifestyles, behaviors or peculiar sex habits that are odd, yet strangely fascinating. There are a seemingly endless variety of curiosities under the sea, and many of them are fish.

AN UNDERSEA COMBO-PACK

Can there be a weirder animal in the oceans than one that combines the head of a horse, the pouch of a kangaroo, a chameleon's swiveling eyes, and the prehensile tail of a monkey? One that also has armor, color-changing skin, and four tiny, translucent fins that beat almost imperceptibly. And in a unique swap of reproductive responsibilities, in this creature it is the male that endures pregnancy's bulge and experiences the agonies of childbirth. The seahorse is an enigmatic creature, and even more so because it is a fish.

Seahorses have fascinated humans for thousands of years. They were once thought to be the offspring of the steeds of the mythical god Poseidon, and later they stymied scientists, who weren't sure how to catalogue them. Were they insects, amphibians, or an odd shrimp-like crustacean? Seahorses eventually became classified as bony fishes, with characteristic swim bladders to aid in buoyancy, and tiny, stiff, and articulated spines, as well as gills for breathing. They are also armored with interlocking plates overlain by skin with pigmented chromatophores that enable color change and camouflage (plate 15).

There are at least thirty-nine species of seahorses recognized to date. They range in size from the less than 2 centimeters-long (inch-long) pygmy seahorse to the big-belly species standing almost 30 centimeters (a foot) tall. Seahorses are variously colored and may sport spikes, knobs, or zebra stripes. They live mainly within coral reefs or seagrass beds or among the roots of mangroves. These very unfishlike fishes are typically solitary creatures that spend most of their time sitting quietly upright with their prehensile, monkey-like tail wrapped around an appropriate hold, such as a blade of seagrass or sponge. But even when sitting quietly, seahorses are actually quite active, constantly scanning their surroundings for food with eyes that can rotate independently of one another, swiveling like a gunner's turret. Their long, toothless, horsey snouts act like powerful slurp guns to suck in the sea's tiny crustaceans, and they can maneuver from one perch to the next or shift position ever so slightly using their small, rapidly beating fins. The seahorses' wispy fins also come into play when dancing seductively for a mate.

The pairing of two seahorses begins with a nose-to-nose greeting and a graceful courtship in which they perform a spiraling ballet about one another. They may also blush colorfully, bow their heads, swim in synchrony, and flirtatiously entwine their tails. The male seahorse may hum gently or make a clicking sound as he attempts to woo a female. It is a dance for the ages that can last for hours, sometimes even days. A female's acceptance is signaled when she points her snout skyward and stretches out straight. In response, the male pumps his tail vigorously, like the gyrations of a Las Vegas performer. When the seduction and preparations are complete, the pair will mate belly-to-belly and snout-to-snout. It is but a brief intimacy as the female seahorse shoots an egg-rich liquid from her body through a short hollow tube into her partner's pouch. They then break apart and the male swims away, though the female may periodically pay him visits. Within the male's pouch, the eggs are fertilized and will incubate for several weeks. As his stomach swells, the male seahorse moves little. The delivery of his offspring is announced by the spasms of labor; it may be a quick birth or take days for all the tiny seahorses to wriggle free from his pouch, with as many as several hundred or even a thousand or so babies emerging. Parental care ends at birth, and the young seahorses are immediately on their own. Many seahorses are monogamous throughout a breeding season and some remain devoted for life.

The seahorses' close relatives are also among the oceans' oddest of characters. The leafy seadragon resembles the seahorse, but in fancy dress, festooned in ribbons. Skinny outgrowths of skin make this creature hard to distinguish from a tangle of seaweed. It swims more horizontally than the seahorse, as does another of its peculiar relatives, the hairy ghost pipefish, which looks like an elongated, stretched-out seahorse densely swathed in skinny streamers, or as biologist Helen Scales puts it, "like a monster from the *Muppet Show*."

ANGLING FOR FOOD

The seahorses seem like practitioners of Zen within the oceans, sitting ever so still, but the true masters of the art are a group of peculiar fishes that have a knack for immobility and a body built for angling. The deep-sea anglerfishes, frogfishes, and their relatives are odd, mouthy creatures that have a built-in fishing pole and lure. The deep-sea anglerfishes are the icons of bizarre, when it comes to life in the oceans' depths, where food is elusive and energy conservation is essential. They are rounded, fleshy, dark-colored fishes that are relatively small and without scales or much muscle. The anglerfishes are not outfitted for the fast or long chase, but are built to wait patiently and draw in their prey. One of their front dorsal spines has been modified into a thin, flexible rod, tipped with an enticing fleshy appendage, or lure. An anglerfish dangles, flicks, or gently waves its bait to mimic a wriggling fish, worm, or crustacean. Some species have lures that are bioluminescent and glow temptingly in the darkness below. Other anglerfishes have tasty-looking appendages that hang from their chins. Like many of the deeper living fishes, they have large, expandable mouths that allow them to eat relatively large-sized prey and reduce the chance that a meal will get away. If the lure of an anglerfish is eaten while fishing, handily it has an unlimited supply, as it can regenerate its bait. In addition to being excellent anglers, these fishes have another claim to fame, in the category of really strange sex.

Female deep-sea anglerfishes spend their days hanging motionless, fishing patiently, waiting for food to come their way. Male anglerfishes, on the other hand, seem to have a very different mission in life. They are more muscular than the females, and seem to have a better sense of smell and vision, but alas they are typically only one-tenth the size. Essentially, they are tiny (about a centimeter long) sperm producers whose only goal in life is to search the vast depths of the sea for a female, seeking out her alluring scent or attractive form. Beauty is definitely in the eye of the beholder in this case. When a dwarf male finds a female, he latches on, and not meta-phorically. He actually bites onto the female anglerfish for a never-ending kiss as his mouth becomes fused with her skin. His internal organs then begin to degenerate, with the exception of those that produce sperm, and he becomes reliant on the female's bloodstream for nutrition. Once mated, the tiny male anglerfish is a literal parasite of the female, living and functioning solely to produce sperm for as long as she lives. Usually, just one male attaches to a single female, but there can be as many as eight "hanger-oners."

Related to the deep-sea anglerfishes are the fascinating, somewhat goofy-looking frogfishes, which are found in relatively shallow water within the tropics and subtropics, often in coral reef habitats. They have globular bodies, short paddle-like tails, small top-set eyes, and really big, upward-opening mouths—if you can discern a mouth. In fact, it is often difficult to distinguish a frogfish from a brightly colored sponge or an algae-covered rock. Their skin can take on a wide range of colors, textures, and mottled patterns that often match or blend in with their surround-ings. Names such as the warty or clown frogfish, the leopard frogfish, and the striped or hairy frogfish provide apt descriptions of their varied garb. They sit on and walk with their bottom pelvic and pectoral fins, the latter of which have an elbow-like bend making them appear very much like feet. Like its relative the anglerfish, the frogfish has a modified dorsal spine that constitutes a rod and lure. Its bait may dangle just above its head or be extended farther out on a long, skinny antenna-like spine. Sitting motionless on the bottom, the frogfish waits patiently for a fish, worm, or crustacean to come close, attracted by the prospect of a tasty meal (plate 16). When a victim is in range, the frogfish lunges with rapid finality. Frogfishes do not have the same extreme gender differences as in the deep-sea anglers, and it is often difficult to tell males from females, except just before spawning, when the females' bellies bulge with eggs.

Another odd member of this clan is the flattened, nearly all-mouth monkfish. It is not a pretty fish by any means and is literally almost all head, with a large, toothy mouth and just a bit of body and tail behind. It lies buried or hidden at the seafloor, just waiting for the chance to gobble up the unwary.

ODDBALL SHARKS

Sharks are among the oceans' top celebrities, with a host of television shows, documentaries, and movies to their credit. They even command a whole week on the Discovery Channel every summer. Sharks have become symbols of the sea and its most predacious of predators. But with more than 350 species of sharks in the oceans, there are many that don't fit the stereotypical "Jaws" image; they are in fact ocean oddballs.

The cookie cutter shark is one of the strangest of its kind, with a notorious and ghoulish penchant for plugs—plugs of flesh that is. This shark lurks principally in the oceans' midwater zone, but may swim to the surface at night or go deeper by day. The cookie cutter is a relatively small shark, growing to lengths of about 45 centimeters (1.5 feet), with a brownish, almost cigar-like, body and a small dorsal fin to the rear. It is well outfitted with photophores and can produce bright bioluminescence. The cookie cutter's notoriety comes from its jaw structure and one really weird strategy for feeding. Using its strong suckering lips, it attaches itself onto the skin of large prey, such as a whale, dolphin, fish, or another shark. Once latched on, the cookie cutter bites in with jaws that are lined by small razor sharp, hooked teeth on top and a band saw on the bottom. It then swivels around its jaws and like a cookie cutter slicing through pastry dough, pulls out a rounded plug of flesh from its victim. When not cookie cutting, these sharks also feed on squids, fishes, and small crustaceans. They gained nefarious fame in the 1970s, when the sonar domes of several American nuclear submarines suffered strange circular-shaped damage. The neoprene coverings of the sonar domes were subsequently replaced with an anti-cookie-cutter-shark version made of hardened fiberglass. You can bet that even the best military minds did not foresee this undersea adversary. Like many sharks, the cookie cutter gives birth to a few live offspring that develop within the female.

The sand tiger shark also delivers live young, but for these pups, training to be a top predator begins exceptionally early—inside the womb. It is sibling rivalry taken to the extreme in the form of intrauterine cannibalism. In a mature sand tiger shark female, multiple embryos develop within each of two uterine chambers. The largest of the embryos in each uterus hatches first, typically when they are at least a centimeter (0.5 inches) long. Already teethed and hungry, the young pups proceed to kill and devour the other hatched or developing embryos within the uterus. Once done with their potential siblings, they attack and consume any unfertilized eggs. The pups are born after an overall gestation period of some nine to twelve months and will be almost a meter (about 3 feet) or so in length. Mature females usually give birth to one or two pups annually or every other year. Even for sharks, which tend to have a low reproduction rate compared to other fishes, the sand tiger shark produces relatively few young. Nonetheless, they have a competitive edge early on; the pups are born relatively big and they are already experienced hunters, having attacked and killed while in the womb.

Sand tiger sharks, not to be confused with tiger sharks, are the scary-looking sharks commonly found in large aquarium displays. They grow to be about 3 meters (10 feet) in length and are gray-green to brownish on top and light underneath. They can be stout sharks that look almost hunched, with elongated tails that have long upper lobes, and two nearly equal-sized triangular dorsal fins. The sand tiger shark's most obvious (and scary-looking) feature lies just below a short, slightly upturned snout—a mouth full of long, ragged, spiked teeth that give it a snaggletoothed grin. It feeds on a wide assortment of fishes as well as squids, crustaceans, and smaller sharks, and is typically found in temperate and subtropical coastal waters, tending to live in shallow bays, sandy areas, or regions of rocky or coral reefs. Despite its fierce appearance, it is not considered especially dangerous unless provoked—or if you happen to be a female of the species and it is mating season. During courtship, the males bite the females and then use their toothy mouths to hold on while inserting one of their two claspers. The claspers are the shark equivalent of a penis and are used to transfer a package of sperm from the male to the female. In some sharks, the claspers are outfitted with hooks or spurs to help keep them in place during mating. The sand tiger shark has one more unusual talent for its kind. It is the only shark known to gulp in a bubble of air at the surface to hold in its stomach for buoyancy. Throughout the world sand tiger sharks go by several aliases, such as the gray nurse shark, spotted ragtooth, ground shark, and slender-toothed shark.

Another strange shark that appears in need of dental work is the goblin shark. Its stiletto-laced jaws, however, are not always on show, as they are loosely hinged and retractable, probably in view mostly when feeding. Goblin sharks have unusually long, soft, and rod-like snouts. They are poorly understood, mysterious creatures with bodies that are soft and flabby and that appear oddly pinkish-gray due to a thin, translucent covering of skin that reveals the underlying blood capillaries. They have two rounded, similarly sized dorsal fins and long ribbony tails. The goblin shark can grow to over 3 meters (10 feet) in length and generally inhabits depths between 40 and 1,200 meters (130 to 3,900 feet). Though their distribution and abundance are not well known, they may be relatively widespread. From the few goblin sharks that have been caught and studied, scientists believe them to be ambush predators that sit and wait for prey such as small midwater fishes, crustaceans, and squids. They may project their jaws and create a powerful suction that hoovers in their victims. In fact, their spiked teeth may be used more like a cage to trap food and prevent its escape rather than to tear it apart; their rear teeth appear well suited for crushing. It is also thought that their long, soft snouts provide for enhanced sensory capabilities.

Like other sharks, the goblin shark is probably a supersensitive creature of the sea, though not in the emotional sense. The acuity of a shark's vision varies with species and depends on where it lives. The goblin shark has relatively small eyes and lives somewhat deep, so it may be sensitive to light, but not have particularly good vision. It has two distinct nostrils below its long snout which probably provide an excellent sense of smell. Researchers have learned that some sharks have scarily good noses and are able to detect certain substances in seawater at concentrations less than one part per million—a teaspoon of blood in a typical-sized swimming pool. The goblin shark, like other sharks, presumably also has taste buds within its mouth that are used to decide if something is actually palatable once bitten—not very convenient if you are the tastee. It may also have good hearing, tactile sensing in its skin, and a lateral line system along its body to sense vibrations or water movement. Sharks also have an unusual sixth sense that enables them to detect weak electric fields or impulses using a system known as the ampullae of Lorenzini. Located about a shark's head, snout, and mouth, it consists of numerous small, hair-lined pores connected to tubes filled with conductive mucus (electric sensing slime) and linked to inner nerve cells. The long, soft snout of the goblin shark may essentially be an ultraelectro sensing wand for the detection of prey.

Perhaps the most extreme headgear can be found on the hammerheads. There are some nine species of hammerhead sharks, and they all have an odd, flattened, bar-shaped head. Their large eyes are located on the outside of their heads' wing-like extensions. Scientists have long debated the purpose of the hammerheads' bizarre headgear, hypothesizing that it enhances electric sensing and sense of smell or serves as a hydrofoil that reduces drag and provides lift, which is important because sharks are negatively buoyant. Recent research suggests that having its eyes on extended head wings not only provides the shark with excellent binocular vision, it also supplies a rearview mirror. Hammerheads dine on fishes, cephalopods, crustaceans, and other sharks, and the great hammerhead has a special fondness for stingrays. It swims just above the seafloor swinging its hammer back and forth, like a treasure hunter sweeping a metal detector over the bottom. The reward it seeks is a tasty stingray buried in the sand. Once a stingray is detected, the great hammerhead uses its headgear to pin it down and prevent escape. The shark then incapacitates the ray by biting off a chunk of its wing and circles in to continue dining. Its penchant for stingrays does, however, have a potentially painful drawback—the barbs. Hammerhead sharks have been found with multiple stingray barbs imbedded in their mouths.

These odd-headed sharks are widely distributed in temperate and tropical waters from shallow depths down to at least about 60 meters (200 feet). The scalloped hammerhead, whose snout is indented and slightly arched, is known to form large aggregations and exhibit surprising behaviors, such as corkscrew swimming, which may play a role in the establishment of a social hierarchy. Hammerheads give birth to active pups and may have as many as forty per litter. In 2001, a strange, seemingly miraculous hammerhead birth occurred at the Omaha Zoo. It was especially peculiar because the female shark involved had not had any contact with a male shark in years. At first scientists thought that the female had used stored sperm to fertilize her eggs, but later it was discovered that it was a rare form of asexual reproduction in sharks. Since that time at least one other "virgin birth" has been documented, in a blacktip shark held in captivity. Some sharks, such as the horn and catsharks, do not give birth to active pups, instead they lay eggs encased in thick, protective cases. These embryos are typically deposited on the seafloor, where they develop until hatching.

The list of oddball sharks is actually quite long. Other examples are the sawshark, whose extended snout resembles a two-sided, spiked saw blade, and the angelshark and wobbegong, whose bodies are strangely pancake-flat. The mottled wobbegong also sports a weird, scraggly mustache of fleshy appendages that hang from its wide upper lip. Another shark, the swell shark, seems like a rather typical small, blotchy-colored shark. That is, until it gulps in water and swells up like a pufferfish to wedge itself into cracks or to appear larger to deter potential predators. But the frilled shark may be the oddest of the odd when it comes to sharks, resembling a wide-mouthed, ruffle-collared eel. It grows to at least 2 meters

(6 feet) long and has one dorsal fin located to its rear near a ribbony tail. Its six gills are frilly-edged, and within its large mouth are three hundred trident-shaped teeth in rows that oddly run from front to back. The frilled shark also has pelvic and anal fins far back on its body reminiscent of the feathered wings of a dart. Little is known about this elusive creature, but it appears to live in relatively deep water and feed mainly on squids and fishes. Japanese researchers studying the reproductive biology of frilled sharks found that they give birth to some six pups at a time and that for the females pregnancy may be an extraordinarily long, drawn-out affair, possibly lasting for over three years.

The size of sharks is quite varied, from the dwarf lanternshark that at maturity may reach only 18 centimeters (7 inches) in length to the largest fish in the sea, the whale shark that can grow to be some 13 meters (45 feet) long. The whale shark, along with its relatives the basking shark and deeper-dwelling megamouth shark, is one of the oceans' enormous but gentle giants. A filter-feeder, it sieves the oceans' waters for plankton, squids, tiny crustaceans, jellyfish, and small fishes. The whale shark is distinctively dressed in white spots and bands that create a checkerboard-like pattern on their blue-gray backs. It has a broad, flat head and a very wide mouth containing thousands of tiny teeth. Whale sharks swim through the sea with their huge mouths open wide or position themselves vertically and suck in enormous quantities of water. Although they are considered plankton feeders, larger fishes have been found in their stomachs, presumably having been accidentally sucked in. The sheer size of the whale shark and its cousins is astonishing given that, like some whales, they gain their behemoth status by feeding relatively low on the food chain.

Hundreds of millions of years of evolution have made sharks some of the most successful animals in the sea. Of the hundreds of species of sharks in the oceans today, only a handful pose a danger to humans. When incidents do occur, most of the time it is a case of mistaken identity and the shark is just being a shark, hunting for a meal—of the nonhuman variety.

Some creatures in the ocean seek out the company of sharks—and whales, billfishes, sea turtles, and dolphins. The remora, sometimes called a sharksucker, attaches itself to hosts by suctioning on with a flat, slatted organ that is actually just the top of its odd head, which develops from a spiny dorsal fin when it is young. Besides free transportation and protection from predators, hitchhiking provides remoras a ready supply of food, from their hosts' feeding scraps, parasites, and as some scientists hypothesize, from their feces as well. They are typically small fishes and may have a thick black racing stripe running along their bodies. Remoras are not all that discriminating, and sometimes can be an unexpected and tickling nuisance to a diver or surprised swimmer.

OTHER ODDITIES

The sea's weird fishes are like a never-ending carnival freak show, with new acts frequently being discovered and old favorites that never get boring. The diamond-shaped manta ray is an odd but majestic relative of the shark, having a body similarly built of cartilage and lacking in bones. Just in front of its eyes, the manta has two distinctive, peculiar fins that jut out from its wide mouth, which are used like a funnel for feeding. Mantas can grow to huge proportions, with wingspans of over 6.7 meters (22 feet). They are eerily dark on top and blotchy on their undersides, with short tails that lack the stinging spine carried by other rays. The manta ray uses its large pectoral fins to fly gracefully through the sea. To feed, the manta may swim upward in a wide spiral, opening its mouth to become a giant sieve, filtering the ocean for food. One of the strangest sights in the sea is when a huge manta ray leaps out of the water, jumping high into the air as if shot from an undersea cannon. Sometimes it does a barrel roll before gravity pulls it back, splashing into the watery depths. Like other rays that jump, they may do this to remove dead skin or parasites, or when being chased, or for play, or maybe as a form of communication or courtship. Mantas and other rays sometimes swim at the surface with the tips of their wing-like fins sticking out; from afar they can easily be mistaken for sharks.

Another peculiar and enormous fish is the silvery-gray *Mola mola*, or ocean sunfish. It can grow monstrously large, some 4 meters (13 feet) across, and top out at thousands of kilograms (pounds)—it is the uncontested heavyweight of the bony fishes. It is also one of the rare fish designs that just seems wrong, more like a giant's thick dinner plate with fins stuck on or a species whose designers quit before finishing the back half. The *Mola mola* has a broad, laterally compressed body with two big eyes, a small mouth, and a blunt, rounded, seemingly cut-off back end. It propels itself oddly by paddling with its two long fins positioned to the rear; one that points up and one pointing down. It also has gritty, sandpapery skin that is coated in mucus—more slime. The *Mola mola* uses its fused, beak-like teeth to feed principally on jellyfish as well as other organisms such as squids, sponges, or crustaceans. It has been seen lying strangely on its side at the surface, presumably sunning itself; hence the alias "ocean sunfish." One female *Mola mola* was found carrying 300 million tiny eggs, a potential record in fish fertility.

Going on bizarre looks alone, a few more examples come to mind. Like its namesake, the pineconefish resembles something that fell off a tree rather than a swimming, breathing creature of the sea. It is relatively small, some 12 centimeters (5 inches) in length, and covered with large, bright yellow scales that are each distinctly outlined in black. It is found in the tropical-to-subtropical region of the Indo-Pacific to depths of about 200 meters (650 feet), often in reef habitats. Another apt name for this unusual fish is the Japanese pineapplefish. The oarfish, sometimes called a ribbonfish or the king of the herrings, is not only another odd fish, but perhaps also the explanation behind numerous sea serpent sightings. It is a long, serpent-like fish that can grow to record lengths, possibly over 6 meters (20 feet). Its highly reflective body is bluish to silvery with a crest of scarlet fin rays running along its back and deep-red fins. Its two name-sake pelvic fins are long, stickish, and tipped by rounded membranes. The oarfish is known to hover oddly in the water with its head up and feeds on small fishes, crustaceans, and squids. They are mysterious creatures, seeming to live within depths to about 910 meters (3,000 feet) or so, and may occur widely throughout the world.

The epitome of an ocean oddity, however, may be the barreleye fish, which comes sporting a see-through head and tubular eyes. It is another wacky wonder that is rarely seen, and is found at depths where only a very faint dim of light shines down from above. In addition to two large nostril-like organs at its front and a small, slightly pointed mouth, the barreleye fish has a bizarre fluid-filled, transparent shield covering its head. Within this see-through dome sit two large eyes that are thought to be good at col-lecting light, but that give it tunnel vision. To compensate for its limited view, the barreleye can rotate its eyes so that it is looking up or straight ahead. Its stocky, soft body and large, flat fins suggest that the barreleye is an ambush predator, hovering at depth, waiting for prey such as jellyfish to near.

In the oceans, it is not only looks that can render a fish peculiar, but also an alternative or changing lifestyle. The flatfishes such as flounder, hali-but or sole, offer an eye-popping example. They begin life looking pretty typical for a fish, with an eye on each side of their heads. But during their larval stage, something very strange happens—one eye migrates around the head to the other side. They then settle to the bottom, lying down on their blind sides. Some flatfish species are right-eyed, while others are left-eyed. The right-eyed halibut lies on its blind left side, while the left-eyed peacock flounder lies on its nonseeing right side.

Among the oceans' many fishes are some that are only temporary residents within the sea, and others that travel abroad for breeding. For example, the Atlantic, chinook, and sockeye salmon are residents of the ocean most of their lives, but then migrate into the freshwater streams and rivers of their birth to spawn. How these organisms navigate to their breeding grounds has long been a mystery and topic of keen scientific interest. Cues for navigation are believed to come from chemical smells, temperature gradients, lighting, and the Earth's magnetic field. Moving in the opposite direction are the American and European eels, which spend their adult lives in freshwater and then swim to the Sargasso Sea to spawn. Surprisingly little is known about their time in the sea or how they too navigate through the oceans' unmarked waters for a sexual rendezvous that will hopefully produce a genetic legacy.

WHY THEY MATTER

Within the sea, fishes are the food that fuels the growth and reproduction of many organisms, including other fishes. As consumers of phytoplankton, zooplankton, and other prey, fishes also help to keep a wide variety of populations in check. No matter how bizarre or commonplace, big or small, shallow or deep, all fishes play an important role within the oceans' web of life. They also have both obvious as well as more subtle connections to humankind.

Food, Health, and the Economy

Over three billion people rely on seafood for a major portion of the protein in their diets, and even more eat fish as part of a healthy lifestyle. Local and global economies are supported by the millions of jobs directly and indirectly related to the fishing and aquaculture industries. Simply put, fish are integral to the economic stability and health of world populations. And it is not just the fishes we consume that are important. We may not dine on a certain species of deep-sea shark or anglerfish, but they may be the food of the species we do eat. We may not eat whale sharks, manta rays, or ocean sunfish, but they may be critical to keeping regional plankton or jellyfish populations under control. Every filter-feeding, grazing, or predatory fish in the sea is part of the complex web of life that allows all species to exist, including the ones we like to put on the dinner plate.

Our insatiable and growing appetite for fish has led to the heavy exploitation of numerous species in the oceans, such as tuna, orange roughy, cod, and swordfish. As these fisheries decline or collapse, we have extended our efforts to fishes that were once thought of as trash or too repulsive to consume. The big-headed, all-mouth monkfish, a relative of the deep-sea anglerfish, was once called the poor man's lobster, but is now a popular, pricey commodity gracing many a gourmand's table. What one culture may consider an ugly oddity or strange beauty, another may make a staple of their diet or much treasured delicacy. In the United States, the great majority of people would never imagine eating a seahorse. In Asia, however, seahorses make for fine snacking, boiled, roasted, skewered, or deep-fried. In China, the consumption of seahorses has a long history, stemming from their use in traditional herbal remedies. Eels are also popular in Asian cuisine as well as in the fare of some European countries, where they are typically enjoyed fried, jellied, or smoked. In some places whale sharks are eaten; in other areas manta rays are consumed. Throughout the world, shark meat is devoured in great quantity, eaten as steaks, fillets, in fishcakes, burgers, and the ever-popular "fish and chips." In Asia, shark fins are used for the high-priced and traditional shark-fin soup. One bowl can sell for as much as $200 dollars. Tens of millions of sharks each year are cruelly slaughtered just for their fins to feed the demand for shark-fin soup. Sharks are also fished to grind up for fishmeal, animal feed, or fertilizer, or for their skin or oil. As traditional fisheries decline, people will undoubtedly turn to other species for food—will jellyfish, deep-sea anglerfish, or the oarfish eventually be the cuisine du jour?

Medicine

For centuries, humans have looked to the sea for compounds to heal wounds, reduce suffering, and cure disease. These remedies are often steeped in myth and tradition, and the unscrupulous may even use them to make an easy buck. In some cases, traditional medicines have proven quite effective and the science underlying their success has been documented. Many herbal remedies, however, remain scientifically unproven. Seahorses have been used in traditional Chinese medicine for thousands of years, to cure anything from broken bones to asthma, ulcers, and a weak libido. In Japan, seahorses are considered an aphrodisiac, and in Vietnam they are used as energy-giving snacks. In Roman times, a mixture that included charred seahorses was touted as a cure for baldness. Some cultures consider these small charismatic creatures good-luck charms or talismans to bring great fortune. Today, millions of seahorses are sold in traditional markets as well as through the more modern convenience of the Internet. Discerning shoppers may, however, be put off by the outrageous claims of ads such as that posted by the Shanghai Pisense Industry and Trading Company claiming its seahorse-penis-containing pills as a cure for impotence and men whose endowments are, well, less than desired. To date, there is no scientific evidence supporting the use of seahorses in medical applications, yet some 25 million are plucked from the oceans each year for the medicinal market as well as the aquarium trade and to be sold as trinkets.

Sharks have also fallen prey to the unproven claims of herbal remedies, from the use of shark bile and gallbladder for eye cataracts to dried shark brains to prevent dental decay or ease the pains of labor. But perhaps the most widely sold and overblown ocean-derived remedy is that based on shark cartilage. A wide assortment of pills and powders containing shark cartilage is marketed to reduce joint pain and inflammation, to provide immune support, and as a source of calcium. Some makers have even claimed it as a cure for cancer. To date, there is no scientific evidence to support any of these claims. In 2000, the U.S. Federal Trade Commission ordered two companies to stop their claims regarding shark cartilage as an anticancer drug and fined one of them a million dollars for false advertising.

In other areas of medical research, however, sharks are proving quite fruitful. For example, high concentrations of an antibiotic compound called squalamine were discovered within the liver of the dogfish shark. Squalamine was later synthesized chemically in a laboratory and is being tested as a treatment for a wide range of diseases. It is showing great promise in the treatment of adult macular degeneration, which is the leading cause of blindness in the Western world and affects tens of millions of people. Squalamine has also been shown to produce antitumor activity, which may aid in therapies for cancer, such as advanced ovarian cancer. In animal studies, squalamine was shown to have a significant effect as an appetite suppressant and is now being studied as a means to combat diabetes and obesity. Sharks are also being used as biomedical models to better understand how kidneys function and respond to diuretics, and to learn more about the rejection of organ transplants. Responsible research using sharks relies principally on species that can be bred in a laboratory and focuses on developing compounds that can be chemically synthesized.

Salmon have proven a boon to human health as well, through more than just a plateful of low-fat fish. Salmon calcitonin is a hormone that has been effectively used to treat osteoporosis, and research has shown that it in humans it is fifty times more effective than the associated natural human hormone.

As we learn more about the oddest of creatures in the sea, it is becoming clear that even they may have potential in biomedical research. Take for example the bizarre sex life of the deep-sea anglerfish. Researchers are studying the parasitic attachment and melding of the dwarf males with the females in regard to the immune system and endocrinology.

Other Uses

Far beyond our physical use of the oceans' odd fishes is a value that is difficult to quantify. When seen at aquariums or on film, strange fishes often evoke an emotional or intellectual response. The cute seahorse brings smiles and delight, while deep-sea anglerfish inspire wonder and contemplation about evolution and biology. We are startled by the fast-lunge reactions of the well-camouflaged frogfish or big-mouthed monkfish. Sand tiger sharks and hammerheads induce curious, fearful stares and may inspire an unwarranted and unfortunate life-long fear of the sea. And the power and grace of a large manta ray is nothing if not mesmerizing. Odd fishes have, in fact, long inspired myth and legend, and may explain some of the sea's most mysterious sightings. In more modern times, the oceans' unusual fishes have become part of popular culture, providing content for numerous books, television programs, and feature films. How do we quantify the value of marine life that can inspire such creativity, contemplation, and bring about an appreciation for the wonders of the natural world?

In reality, we know very little about the great majority of fishes in the sea. From what we do know, several things are clear. Hundreds of millions of years of evolution have led to an astonishing diversity of fishes that we are only beginning to recognize and to understand. Every fish, no matter how small, big, or peculiar, plays an important role in the oceans' complex web of life, which in turn supports and provides for humankind. People can benefit greatly from fishes through more than just a plate of food. And if we are to take full advantage of the oceans, it will mean protecting the sea as a priceless cabinet of curiosities, too valuable to lose, too vulnerable to neglect.

7 X-Games

With millions of avid fans and dollars in sponsorships, extreme sports have become a global phenomenon, yet in the sea, organisms have long been performing exceptional feats of daring and pushing the boundaries. They have been diving deeper and swimming longer and faster than we can ever hope to. In the oceans, radical matches in boxing and wrestling are regular occurrences, and multisport events continually test the sea's most elite athletes. Extreme action is the stuff of everyday life in the oceans, and the "X-Games" have been going on for ages. Let the games begin, or more appropriately—continue.

DEEP DIVERS

The deep sea is intensely cold and dark, and the pressures are crushing; yet many marine organisms make dives every day into these extremes. Some air-breathing creatures take the plunge on just one lungful of air; the champions among them are the sperm and beaked whales. They have been documented going to the astonishing depths of nearly 2,000 meters (6,560 feet) and staying underwater for as long as one to two hours—all on one lungful of air! One sperm whale may have descended to 3,000 meters (9,800 feet), and some scientists speculate that they can go even deeper. In this ocean x-game, the runners-up are elephant seals, diving to over 1,500 meters (some 5,000 feet) for as long as two hours.

Most animals that dive into the sea's extremes do it in pursuit of tasty, energy-laden fare, such as squids. Marine mammals are especially well suited for the deep-sea hunt. They have blubber to protect them from the cold and a physiology that allows their lungs to collapse at high pressures.

They can efficiently store and use oxygen, and redirect blood flow to their vital organs, the heart and brain. They can also slow down their heart rates and glide to use less energy. Deep-diving seals use their acute eyesight and possibly sense of touch to capture prey at depth, whereas whales use echolocation. By emitting pulses of sound and receiving the echoes that bounce back, they somehow create a detailed picture of their surroundings and the objects in it, somewhat like our less advanced version, sonar.

Pilot whales do not dive to the greatest of depths, but they have speed—they have been called the cheetahs of the deep. As they descend, they echolocate, presumably in search of prey. Engineer Mark Johnson from Woods Hole Oceanographic Institution likens it to "acoustic window shopping." When a fast-paced squid or fish is detected, the chase is on, and a high-speed pursuit ensues. Pilot whales can make short downward sprints at up to 32 kilometers per hour (20 miles per hour).

Our ability to track animals into the deep sea has improved dramatically in recent years, especially with the advent of noninvasive tagging technology. One can only wonder what new information will be unveiled as we learn how deep and how far marine animals can actually dive and just what they do, once they take the plunge.

ENDURANCE ATHLETES

Whales also hold the title in the oceans' ultra marathon, a test of will and endurance over extreme distances. On the podium's top step is the Pacific gray whale, which makes the longest regular migrations of any air-breathing animal on the planet—some 19,000 kilometers (12,000 miles) each year. These 12-meter (40-foot) long, gunmetal gray behemoths swim extreme distances, driven by hunger, lust, and the instinct to reproduce and care for their young. In the summer, Pacific gray whales are found in the Bering and Arctic Seas. The days are long and sunlit, the sea ice has melted, and food is abundant. Blooms of phytoplankton turn into a bountiful buffet for zooplankton, which in turn are feasted on by larger organisms. It is this sea of plenty that lures the gray whales. Of the filtering baleen whales, they are the only ones that feed principally at the bottom. They dive down, roll on their sides, and suck in large quantities of sand, mud, and water. Using their tongues, the gray whales then push the sediment and water through strainer-like plates of keratin (baleen) that line their upper jaws. Crustaceans, worms, and other organisms that are caught on the baleen are licked off and swallowed. From about June to August, the gray whales gorge themselves, adding layers of blubber to their enormous bodies. They bulk up for warmth and to store energy for the long, foodless days ahead.

The gray whales leave their northern feeding grounds in the fall and head south to Mexico. Over the next several months they travel thousands of kilometers, and though they may intermittently feed, it provides little sustenance. As they migrate, the gray whales are often seen spyhopping off the western coast of the United States, bobbing up and looking around, presumably to identify landmarks that help guide them south. Along the way, mating may occur, especially as they near their warm-water destination. Once in the shallow, warm, and protected waters of Baja California, the gray whales rest, and those that are pregnant give birth. They will spend the next several months nourishing their young with fat-rich milk. Whale milk, aka a blubber shake, is up to 40 percent fat, and calves may consume up to about 190 liters (50 gallons) daily, allowing them to pack on the pounds. The young gray whales' appearance has led some people to call them "pickleheads." The gray whales of Mexico's lagoons have acquired another nickname as well, "friendlies," because over the years, they have begun inexplicably to interact with whale watchers in small boats, seeming to even seek out human touch and interaction. After about two to three months in Mexico with little to no food (except the calves of course), the gray whales leave their extreme blubber-beating boot camp to begin the long return journey north. The mothers and calves are the last to leave. Without a truly decent meal in months, they undertake another ultra marathon swim, heading back toward the Arctic's cold, food-rich waters.

Humpback whales are the runners-up in this ocean x-game, but they are by no means slackers. Found worldwide, most humpbacks annually migrate from relatively high-latitude locations in the summer to the tropics or subtropics in the winter. Humpbacks are easily recognized by their distinctively bumpy heads and long flippers, which extend to about one-third of their body length. Schooling fishes and krill provide the fuel for their success as endurance athletes. Humpbacks often forage alone, but sometimes they will work as a team in a unique and ingenious means of capturing prey. When a pod of humpbacks comes upon a school of fish, one may dive down below and begin producing bubbles. The bubble-blower creates an ascending corral of air that frightens the fish and causes them to tighten their ranks and move toward the surface. Other members of the pod may herd the fish physically or through distinct feeding calls.

Then, in coordination, the humpbacks all dive down beneath the bubble-netted fish and lunge upward with their mouths opened wide (plate 17). It is a team effort that concentrates their prey and ends with a rich and tasty group reward.

Humpbacks summer in locations throughout the world, including off the coast of Norway or Iceland, in the Gulf of Maine, the North Pacific, or in the Antarctic. Come fall they head to warmer waters, where, like the gray whales, they breed and calve. While the rest of the group travels, some nonbreeding females may remain in their feeding areas. In the Arabian Sea there is an anomalous population of homebody humpbacks, none of which migrate.

Humpback whales are also the showboaters of the sea, with an aptitude for aerial performance and a talent for song. In a breach they leap head-first, propelling their enormous bodies out of the water for some airtime. They may also slap their tails or flippers and spyhop. Why humpbacks perform such acrobatic feats remains a matter of speculation; tail slapping may be a sign of aggression or harassment, and breaching a means to clean off sloughed skin and parasites. These behaviors may also have other social purposes or play a role in communication. The truth is that while we ooh and aah over whales and their behaviors, we can only guess at their reasons.

Male humpbacks dominate the sea in song; they are the uncontested winners of the world-ocean idol competition. They sing plaintive, sonorous songs that intermix deeply resonant tones with high-pitched and pulsed notes. Within each of the world's ocean basins, the male humpbacks sing a different song. It is always the males that sing, and over time their songs may change or progress. Amazingly, all the males within a region sing the same song even as it changes. The humpback's ballad appears to be a vocal want ad for sex, to attract a mate. They may also use it in competition with other males. At their summer feeding grounds, male humpbacks interact and may forage together. But when mating season arrives, the flippers come up. The males get competitive, and not just with song—they can get physically violent. In this society, however, it is the females that do the choosing, and not all males prove worthy. Sometimes strangers with a novel tune may come to call and become singing sensations with great influence over the locals. In 1996, a few distinctly new and different songs were heard within a population of male humpbacks that migrate along the eastern coast of Australia. The novel tune was the same song that western Australia humpbacks sing. Presumably, a few long, dark strangers from the western population were visiting the humpbacks in the east. The eastern population of humpbacks began to take up the strangers' tune, and after two years, all the males in the area were singing the new song.

Some may sing love songs, but male whales do not appear to be the most loyal of mates. There seem to be no monogamous whales; they are either promiscuous or polygamous. Male gray whales and dolphins are indiscriminating and have many partners, while sperm, humpback, and orca males prefer to mate within a specific group, but still like to have multiple wives. The right and bowhead whales have their own claim to reproductive fame. The males of these species do not seem to challenge each other for the affections of the females; instead it is their sperm that does the competing. Once inside a female, the males' small swimmers battle it out or race to be the seed that fertilizes her egg. In fact, right whales have been observed in ménage à trois: since there is no competition between males, two may mate with the same female at the same time. Evolutionarily, for these whale species the more competitors (sperm) they produce and the farther ahead their representatives begin in the race, the better—hence, the advantage of having large storage and long delivery "tools." Relative to their body size, these whales have extremely large reproductive organs, some 13 to 14 percent of their body size. The right whale's sexual sword may be 3 meters (almost 10 feet) long!

And now, for an interesting and slimy side note about whales. The clearing of a whale's blowhole at the surface is a wonderful spouting telltale of their presence. It is also a spewing of oily air, seawater, and mucus coming from their lungs—whale slime. Trust me, not a pleasant combination to be showered with. There may, however, be some very valuable information contained in spouted whale boogers. It just takes a bit of ingenuity to collect them. Using a pole outfitted with a collecting device made from knee-high stockings and extended from a small boat, Australian scientists captured the blow from humpback and northern right whales. Within the whales' mucus they were able to detect reproductive hormones, which may supply an innovative and noninvasive way to assess breeding status and rates. In California, other scientists cleverly used a small, remotely controlled helicopter to collect blow from gray whales. They were able to position the helicopter overhead just as the whales surfaced and cleared their blowholes. These researchers discovered distinct bacterial communities within the whales' mucus that may provide a new means to assess their overall health.

There is one more migration in the ocean that is worthy of a place on the x-games' winners' stage. Some scientists consider it the largest synchronized movement of biomass on the planet. It is the nightly passage of small fishes and zooplankton from the sea's depths into shallower water. For some of these tiny organisms, the upward distance traveled is equivalent to fifty thousand times their body length. In human terms, that would be like an evening stroll of some 80 kilometers (50 miles). Though not fully understood, it is believed that the sea's small zooplankton and fishes move upward at night to forage at the food-rich surface at a time when visual predators are at a disadvantage. In addition, most of the creatures involved are cold-blooded; the surrounding water regulates their temperature and metabolism. Staying in deep, cold waters during the warmth of the day and migrating upward at night may help them to conserve energy. The nighttime sky appears key in controlling this mass vertical migration, and when nights are especially dark, without a moon or heavily overcast, the small creatures of the sea rise up in force. In contrast, on bright moonlit nights, vertical migration is subdued. This daily upward marathon is well studied in the tropics, but is just beginning to be documented and understood in the planet's colder realms.

SPEED FREAKS

Open ocean predators often feed on fast-swimming fishes or squids, so they have a need for speed. Certain traits are advantageous if one wants to be among the sea's top sprinters; these include being torpedo-shaped and streamlined, and having a stiff, forked tail that provides powerful propulsion. The swiftest of the oceans' athletes are also outfitted with fins that can fold into specialized grooves on their bodies and drag-reducing hydrofoils in the form of finlets or tail keels. Being pumped up with red muscle enables a fish to sustain high speeds, while having more white muscle allows for quick bursts of energy. Some of the oceans' speedsters also use special breathing tactics, a phenomenon called ram ventilation. They swim with their mouths open so that water is literally rammed in and over their large gills.

As in other elite sports, the competition for the world's fastest ocean swimmer is steeped in controversy. Much of the debate stems from the problem of just how to compare the speeds of animals in the open ocean—it is not as if you can call, "On your mark!" and fire a starting gun. Reported speeds are often anecdotal, or they may be estimates based on models or on the rate at which fishing line is spooled out, or refer to leaping rather than actual in-the-water motion. The top challengers in this contest appear to be the billfishes, such as sailfish, swordfish, and marlin, all of which have been clocked leaping at speeds of more than 80 kilometers per hour (50 miles per hour). Their modified upper jaws, which extend forward as water-splicing bills, and their head shape are thought to provide a competitive advantage. The tuna and wahoo are also up for the title, having been recorded leaping at speeds of more than 64 kilometers per hour (40 miles per hour). The shortfin mako shark and killer whale can kick up some speed as well, on the order of 48 kilometers per hour (30 miles per hour) or more. At slightly lower speeds are the small Dall's porpoise, along with the blue and pilot whales. For style and use of slime, an honorable mention goes to the great barracuda, which can apparently attain speeds of more than 32 kilometers per hour (20 miles per hour). Glands beneath its scales secrete a sleek sheath of mucus that is believed to reduce drag by some 60 percent.

In the oceans there is another organism that exhibits amazing speed; it is one of the fastest of its kind. It has no fins or tail and does not hunt down its prey, but nonetheless its speed provides a distinct competitive advantage. In its first year of life, this organism grows to enormous lengths and successfully combats the sea's powerful waves. It can outcompete all of its challengers and at the same time create a habitat that is home to thousands of other creatures.

Kelp is the fastest-growing seaweed on the planet, with some 100 species occurring worldwide. In just one day, bull kelp can grow as much as 25 centimeters (almost a foot). Typically found along rocky shores in relatively cool, nutrient-rich waters, kelp has recently been discovered in deeper waters in more tropical habitats. It is a brown alga that most often lives fixed to the bottom by a disk-shaped or fingered holdfast or attachment. It has a stem or stipe, and blades that are the ocean equivalent of leaves. The feather boa and sea palm kelps resemble their namesakes, but the two most widely recognized species are probably the bull and giant kelps. The whip-like bull kelp has an amazingly long stipe, growing up to 36 meters (118 feet) in length. At its tip is a large bulbous float from which extend long, yellowy-brown smooth blades. The float keeps the plant's sun-seeking blades at the sea surface and amazingly can contain up to 3 liters (3 quarts) of gas, including carbon monoxide. Scientist Louis Druehl from Simon Fraser University describes the bull kelp as resembling

a "very large gothic brown onion of extraterrestrial origin." While teaching at the University of Alaska, Druehl shared space with a circus elephant
named Bo. One day he decided to do a little experiment and presented Bo
with a large, fresh bull kelp. The elephant enthusiastically grabbed the
kelp and ate the blades. Before eating the float, however, Bo promptly
stomped on it, releasing the gas within. Did the elephant know that the
gas in the float was potentially poisonous based on some distant past experience with kelp?

Amazingly, bull kelp reaches its full length within just one year, growing
from spring into the fall—talk about a growth spurt. Bull kelp is an annual
plant, so that come winter, it dies. The plant material is then broken up and
may be washed ashore or caught in flows that carry it out to sea, thereby
exporting nutrient-rich organic matter to other habitats. Live bull kelp is
also extremely elastic and absorbs the drag of the oceans' waves by stretching up to almost a third of its length, much like polypropylene line.

Giant kelp is the largest species and may reach up to 55 meters (180 feet)
in length. It has a long stem with blades that branch off along its length.
A gas-filled float grows at the junction between each blade and the stem,
with the exception of the first few blades near its base. It too is an ultrafast
grower, and within its first year, the giant kelp may attain a length of up to
about 30 meters (100 feet). Unlike the bull kelp, individuals of this species
can live for up to seven years or more.

Kelp is not, however, the simple seaweed it seems—it leads a double
life. In its more familiar morphology, this speedy grower towers within
the sea and its blades extend to the surface to soak in nutrients and capture sunlight. In this form, kelp releases swarms of kidney-bean-shaped
spores that swim to the bottom, settle, and begin to grow. Hidden from
view, this is the kelp's secret sexy side. The spores become microscopic
plants that produce kelp sex cells, or gametes. Tiny female plants produce
eggs, while miniature male plants produce sperm. The female plant then
exudes a seductive perfume, eau de kelp, which attracts the sperm. Once
fertilized, the eggs begin to swell, change, and eventually grow into the
next generation of the seaweed's other, larger form. Kelp alternates generations between a mega-fast-growing seaweed and a microscopic producer
of sex cells.

Forests of kelp are unique, shadowy worlds rich with marine life. The
dense stands of kelp within create a canopy and habitat that provides protection from predators, a relative calm amid waves and currents, and an
extensive menu for hungry ocean diners. In the kelp beds of the North

Pacific, among thousands of residents there is one cuddly and very hungry celebrity, the sea otter. It uses the kelp like a security blanket while sleeping and as a nursery area for pups. In the kelp forest there is ample food for the otter, including such tasty treats as sea urchins, abalone, crabs, and clams. And sea otters have quite an appetite, a definite challenger in the ocean x-game of competitive eating. To stay warm and keep up their energy, sea otters must consume up to one-fourth of their body weight daily, the human equivalent of eating one hundred hamburgers a day. Still, they stay quite svelte beneath their thick fur; which must be groomed often to add air and coat with oil. Along with being a champion eater, the otter is also one of the oceans' best backfloaters. While floating on its back, the sea otter can sleep, play, groom, or open a clam by pounding it against a rock balanced on its chest. In addition to the famed sea otter, kelp forests provide a home to hundreds of fishes—including one of the oceans' most vibrant and photographed species, the bright orange garibaldi—as well as a myriad of sponges, sea anemones, sea stars, crabs, sea slugs, and more. It is a shadowy and rich seascape dense with life, born by the high-speed growth of the towering and majestic kelp.

THE PREDATOR PENTATHLON

To compete in high-level multisport events, the world's best human athletes need speed, power, and endurance. In the oceans, the ultimate test of athletic performance may be the ability to hunt a wide variety of prey under diverse circumstances—the pentathlon of predation. The top dog in this competition is actually a dolphin. The orca or killer whale is the largest of the small, toothed whales, also known as dolphins. With their high-contrast and camouflaging countershading, black on top and white underneath, along with a tall, relatively narrow dorsal fin, the killer whale stands out in a crowd. It preys on just about anything the sea has to offer, with the exception of humans. Killer whales are fast enough to chase down high-speed fishes and squids. They are smart enough to outwit clever octopuses, dolphins, and other whales. They are also powerful enough to jump onto ice flows to capture seals or to beach themselves on the sand to claim sea lion pups. Orcas have been known to eat sea birds, sea turtles, sharks, and sea otters, and within their stomachs scientists have even found the remains of deer, moose, and pigs. Though there is great diversity in the food that killer whales eat, many individual populations specialize in their prey. For example, in the temperate waters of the Northeast Pacific, salmon are the killer whales' meal of choice; some in the area have an even more discerning palate and preferentially prey on the fattest species, the chinook. Killer whales seem to recognize the value of working as a team and are known to hunt collaboratively and share their catch. They use a variety of tactics to capture, stun, and kill their victims, including head butting, slapping with their tails, or even creating waves to wash seals off an ice floe. Through teaching, killer whales also pass on their hunting skills to their offspring.

Killer whales are cosmopolitan in their distribution, found through-out the world's oceans, though they tend to concentrate in colder regions where productivity is high and food is plentiful. Like a lion waiting in the tall grass near a watering hole on the plains of Africa, the killer whale feeds where its prey congregate. Some killer whales migrate seasonally, while others stay year-round in specific areas. Like other whales and dolphins, they typically use echolocation to detect their prey. In some cases, how-ever, where stealth is an advantage, they will listen passively and hunt in silence. They have excellent gymnastic skills as well, exhibited in breach-ing, spyhopping, wake riding, and doing headstands. In the sea's pentath-lon of predation the killer whale wins by a large margin.

If the oceans' pentathlon were to include the high and long jump, other animal athletes might surpass the killer whale in these individual events. Sailfish, marlin, tuna, and the mako shark are well known for their high-flying leaps. Manta and eagle rays also make frequent jumps into the air. In South Africa, even great white sharks get into the act, with a flair for the high jump. But dolphins too are vying for the title and some penguins might be contenders as well. In the long-jump competition, it could be a battle between salmon and flying fish. Salmon make long leaps upstream as they migrate from the ocean into the rivers of their birth to spawn. Their competitors, the flying fish, regularly make lengthy high-speed glides over the sea surface. Propelling themselves out of the water, they use their outstretched fins like stiff wings and their tails as rudders. In 2008, a Japa-nese television crew recorded a flying fish airborne for some 45 seconds. This possibly record-setting flight was filmed alongside a ferry motoring at about 30 kilometers per hour (20 miles per hour). Flying fish are also known to make unexpected shipboard visits, even entering through open portholes. I have personally seen one fly out of a wave and into the face of a very surprised colleague (it was quite funny for one of us).

IN THE BOXING RING

Within the sediments and rocks of coral reefs lurks an aggressive crustacean that packs a powerful punch. Its odd beauty, small size, and seemingly ungainly appendages hide a mastery of pugilism. The stomatopod or mantis shrimp is a prizefighter able to strike with lightning-quick speed and the force of a 22-caliber bullet, killing an opponent with a single blow. There are some five hundred species of mantis shrimp, many of which live within the relatively shallow waters of the Indo-Pacific and Australia, though they can be found in other regions. Ranging in size from about 2 to 30 centimeters (0.75 inch to 1 foot), they look like a well-armored version of their insect namesake, the praying mantis. Some are particularly colorful and have protruding, stalked, jewel-like eyes in iridescent hues of emerald, ruby, or sapphire. The stomatopod has a hard carapace, or exoskeleton, and an elongated and segmented abdomen, along with sensory antennae and eight legs. Its most formidable and conspicuous attribute is its dangerous second leg, which has been modified into either a spine-covered spear or hammer-like club. Spear-bearing stomatopods live in burrows, lying in wait for soft-bodied prey such as fishes, worms, or shrimp. Quicker than the blink of an eye, a spearer can unfurl and thrust its spine-covered dagger, some out to half their body length. The club-bearing stomatopods, or smashers, are equally as quick and just as deadly, and hunt preferentially for crabs, snails, and clams. Smashers tend to live within crevices amid rock and rubble, and then stalk their prey for a beatdown, pummeling them with the hardened elbow of their clubbed claw. The first punch of a smasher is the stunner, and then it literally batters its opponent to bits. Once either shish-kabobbed or well minced, the mantis shrimp's prey is dragged back to its lair, where the soft tissues are dissected and devoured.

The power of the stomatopod's strike comes from a latch-and-spring mechanism in its claw that enables extremely fast action with enormous force. It takes a camera with a shutter speed upward of five thousand frames per second to capture the mantis shrimp in action. It strikes so fast it creates a cavitation bubble, a flash of light, and a pressure wave that produces a snapping sound. It is believed to be one of the fastest, most powerful strikes in the animal kingdom. Stomatopod lovers beware, as the force of their blow is strong enough to cause serious finger damage or break a double-walled glass aquarium.

The stomatopods also use their weaponry and boxing skills to fiercely defend their living space, though some are more aggressive than others. When faced with an impending fight, some mantis shrimp will attempt to appear larger and more formidable by posturing their antennules, spreading their claws, and putting their awesome armament on display. They also have a brightly colored false eyespot on their clawed append-age, which in some species is distinctive. When displayed, the colored spot may be a coded warning, signaling that a stomatopod is an aggres-sive species or one that has a lot of wins under its belt—either way, not to be messed with. But if a territorial war does come to blows, mantis shrimp may curl up on their backs and use their thickened tail fins as a shield. And if a striking claw becomes seriously damaged in the heat of battle, the stomatopod must physically tear it from its body and wait a few molts for it to regenerate. Molting refurbishes the mantis shrimp's weaponry, but it also makes the creature temporarily vulnerable. To use its smashing or spearing claw when fresh from a molt would mean the risk of literally ripping apart. Rather than reveal their postmolt vulnerability, some sto-matopods will try to bluff their way out of fights by putting on impressive displays as if they are battle-ready.

The mantis shrimp also has extremely keen and complex eyesight, which undoubtedly aids in its pugilistic endeavors. Its movable, stalked eyes rotate and supply excellent color vision. Some species can even see ul-traviolet, infrared, or polarized light (especially good for living underwa-ter in areas bathed in bright sunlight). Color may be used by stomatopods as a means of communication and to signal their sexual desires or mood. Some species may even fluoresce during courtship, truly lighting up for their mates. The mantis shrimp, like many of the oceans' creatures, is of-ten a promiscuous suitor, though a few species are monogamous. In these rare breeds, a pair may live together in a burrow for up to twenty years.

In the oceans, going just one round with a stomatopod is a game of high stakes. These small, super swift, and powerful hunters are among the sea's elite athletes, and their boxing prowess has made them deadly predators.

WHY THEY MATTER

The x-games of the sea showcase marine organisms that have evolved an exceptional aptitude for deep diving, fast swimming, long migrations, and predation. These abilities lend them unique and important roles within marine ecosystems. Top predators, such as killer whales, tuna, billfishes, and the feisty stomatopods, keep prey populations under control, cull the weak or sick, and sustain balance within the food web. Alterations in the population size of such apex hunters or changes in their behavior can have significant impacts on a region. For example, in western Alaska when orcas began killing more sea otters, possibly in response to a reduction in the numbers of their other prey (stellar sea lions and harbor seals), it caused a dramatic increase in the local population of sea urchins, the otters' prey. Sea urchins then began to overgraze and damage the region's kelp forests, illustrating how changes in a top predator's behavior can have effects that cascade through the food web. The oceans' top strainers, such as the gray and humpback whales, are also critical components of the marine environment. These megafilterers help to keep plankton populations in check. But the champions in the oceans' x-games are not just important within the sea; they have strong links to society as well.

Food and Health

For many of the oceans' athletes, what makes them so good at swimming fast or jumping high is also what makes them fine dining. We are eating the oceans' best of the best, breakfasting on its champions. The muscular build of some of the oceans' swiftest, such as swordfish and tuna, puts them at the top of the menu. The tuna's high fat content makes them even more desirable, especially for sushi or sashimi; just one bluefin tuna can fetch a hundred thousand dollars at auction. Worldwide, the tuna and billfish industry are annually worth some five to ten billion dollars and account for 9 percent of the global fish trade. The long-jumping salmon are one of the most popular seafood items in the world, and in some places, flying fish make for a good plate of fish and chips. Not even the mantis shrimp, the oceans' boxer extraordinaire, can evade the dinner plate. Supposedly, they taste a bit like lobster. In several areas of the world, whale meat and blubber are still eaten, in some cases as part of a local tradition or as a means of subsistence. For most people of the world, however, whale meat is thankfully a thing of the past, gone with the days of commercial slaughter.

Kelp has historically been and continues to be harvested and grown for food. Powdered, pickled, dried, made into chips or noodles, and as an ingredient of soup, stews, and salads, kelp is a popular food item in Asia. It is a rich source of fiber, iron, and zinc, is said to have more protein than cabbage, and comes with its own flavor-enhancing salt. High in iodine, kelp provides an important supplement and in a powdered form is suggested to treat thyroid problems and other ailments. Research also suggests that the natural fiber found in kelp may reduce fat absorption when consumed within foods, such as bread or yogurt, and it is now being studied as a potential food additive to help tackle the growing problem of obesity.

Algin is a kelp-derived substance that is commonly used as an emulsifier or stabilizer in foods such as ice cream, milk shakes, salad dressings, baking products, and even beer. That nice head of foam on your next cold beer and its clarity—thank kelp. Kelp is used as feed for farmed abalone and red sea urchins, and in the wild supports a diversity of fishes and other organisms that are collected for consumption. Kelp is also helping to create another high-priced and savored fare—herring roe on kelp, which is formed when herring release their eggs and they stick to and coat flypaper-like kelp in the surrounding water. Roe on kelp is a delicacy, especially in Japan, and is growing as an aquaculture industry.

Fortunately, if managed properly, kelp can be harvested so that it is a sustainable crop, though the same cannot be said for most of the oceans' top predators and elite athletes.

Economy

There is a revolution afoot, or more appropriately a revelation. People across the globe are realizing that the oceans' champions are more valuable alive than dead and that the global economy can benefit from these magnificent creatures more significantly when they remain in the sea instead of on the table. This is no better exemplified than by the shift from the global massacre once called commercial whaling to the huge and growing industry of whale watching. In 2009, it was estimated that some thirteen million people worldwide engaged in whale watching, supporting an industry now worth $2.1 billion. Over the past decade, whale watching has grown dramatically, outpacing the growth of global tourism and bringing jobs and revenue to communities in some 119 countries, providing essential income in some areas that were previously impoverished with little other economic potential. When done responsibly, whale watching can be the basis of a long-term and sustainable tourism industry with broad and positive impacts in coastal communities across the globe.

Drawn by their power, high-flying leaps, and speed, billfishes have long been the obsession of anglers. For years "Catch and kill" was the name of the game, with mounted trophies the reward. Today, many if not most sportfishers practice catch-and-release, mounting only replicas based on photographs and measurements. In Los Cabos, Mexico, the benefits of a healthy billfishing industry were recently revealed in an economic evaluation. The study found that in 2007, more than three hundred thousand visitors fished in the area, spending over $600 million in retail sales, supporting over twenty thousand jobs, and bringing more than $1 billion in overall economic activity. Clearly, these fish are more valuable alive and jumping than mounted on the rec room wall!

Industry

Before many of the advancements brought on by modern technology, marine organisms provided essential materials used in manufacturing and industry. Whale oil and blubber rendered into oil provided an important lubricant and was used in lamps, soap, and in the production of leather and cosmetics. Baleen was harvested from whales for use in combs, parasols, and even ladies' corsets. Ambergris was (and is) one of the most precious (and disgusting) whale-derived substances used in society. It is a gray, waxy substance that builds up within the intestines of sperm whales as a coating on squid beaks that have been consumed. The ambergris comes out of the whale the old-fashioned, natural way. It is defecated as viscous, black, stinky blocks. This ambergris, aka whale poop, is then collected floating in the ocean or washed up on beaches, after it has been worn by the air, sea, and sun. Like a fine wine, it supposedly gets better with age. Depends, I guess, on your definition of "better." Historically, the earthy-smelling ambergris was used as an odor enhancer in perfumes and is still used in fragrances in some areas of the world. In a powdered form it has been used in food or as a supplement. In the Middle East, ambergris is highly valued for its supposed power to promote virility and as an aphrodisiac. But even if no whales are killed, eating or smelling like ambergris still seems wrong, on just so many levels.

Kelp also has a history in manufacturing. In the early 1900s, along the California coast, kelp was harvested as a source of potash, which was used to make gunpowder during World War I. Kelp and its products have also been used as fertilizer, in feed for livestock, and in the manufacturing of cosmetics, lotions, and food products. With the upswing of spas touting the benefits of natural products, kelp wraps and mud baths are now all the rage. Farmed and wild harvested kelp used in food production and manufacturing has an estimated annual value of more than $270 million.

Today, we have synthetic substances that can be used in place of those once obtained from whales, and we have found ways to harvest kelp so that it is a renewable resource. We are also learning that from the very biology that makes our elite ocean athletes elite, we can obtain information of immense practical value. Biomechanical engineers are looking at the design of the humpback's flippers to improve the performance and maneuverability of submersibles. Aerodynamic engineers are examining their design to create better, more efficient fan blades and wind turbines. The stomatopod's eyes are providing insight into complex visual processing that could be used in artificial imaging and vision in robots. The remarkable latch-and-spring action of the mantis shrimp's claw, which results not only in speed, but enormous power, is also being studied for its potential applications. Scientists are investigating how blubbery dolphins and whales can produce so much power and speed within seawater, a medium much denser than air. One interesting finding is that dolphins produce an amazing amount of elastic spring from the stiffness of their blubber and the tissue connecting it to their tails—providing their own in-ocean pogo sticks. And while deep-diving sea lions and dolphins have long been trained for military operations, such as in the detection of undersea enemy invaders or armaments, now tuna are getting into the act. Researchers are designing the "robotuna" to improve the performance of submarines and autonomous underwater vehicles for applications in science, exploration, and defense. Clearly, there is a tremendous amount of knowledge and benefit to be gained by studying what Nature has perfected in the sea, after hundreds of millions of years of evolution.

Culture

From myth and aboriginal tradition to fine literature, popular television or movies and theme parks, the oceans' elite athletes have long evoked our respect, appreciation, and awe. Through Moby Dick, the great white whale has forever become a part of our cultural history. Swallowing-by-whale is now a fictional phenomenon that crosses folklore and popular culture. In some parts of the world, whales and dolphins are revered as part of a local custom or tradition, and many people feel an innate connection to them. We pay millions of dollars each year simply to get a glimpse of these animals. We re-create kelp forests in aquariums and travel to dive or kayak within the coastal ecosystems they create. Some brave individuals even try to keep stomatopods in glass tanks. Artists render billfishes, whales, and dolphins in paint and stone to be purchased and adored. Some people believe that swimming with dolphins even has therapeutic powers. What is it about these creatures and environments that so interests us? Is it simply a whale's size or the mystery of where and what they do, day in and day out? Is it because dolphins have a high level of intelligence and we crave to understand and communicate with them? Or is it something else, difficult to express, which makes our hearts sing when we see a dolphin jumping or a whale breaching or when we gaze upon the shadowy beauty amid the kelp? Many people and communities have moved away from the culture of killing the oceans' top predators and celebrating it as a conquering of nature, but there is still work to be done and those to convince that this change is beneficial to our economy, the environment, and society as a whole.

Every day in the sea there is extreme action. Ultra marathons are swum, dives are made to great depths, and predatory champions are hunting down prey. Kelp is growing at incredible speeds and stomatopods are punching out their opponents. Within the oceans, the x-games are a daily affair. All of the sea's elite athletes play an important role in the oceans and, as we are coming to realize, are worth much more to society alive and abundant rather than dead.

8 Radical Living

Even in the oceans' most extreme, seemingly inhospitable conditions, life not only survives, it thrives. From the frigid cold of polar waters to the hot, acidic seawater at deep-sea hydrothermal vents, life exists in stunning abundance. The assembly of amazing characters in these radical environments includes a small, tuxedoed movie star, a relatively new and fuzzy celebrity, and a bizarre tube-enclosed creature with what is undoubtedly the best name yet—bone-eating zombie snot worm. As humans explore more of the oceans' uncharted waters, new and incredibly peculiar forms of life are being discovered with strange adaptations to radical living conditions. We look to these environments and the creatures living there to better understand life on Earth and its ability to flourish, even in the most hostile of settings.

INTO THE FREEZER

Some of the Earth's most extreme conditions occur within the Arctic and Southern Oceans. In winter, sea surface temperatures can fall to an icy −2°C (28°F), and air temperatures may plummet to below −50°C (−58°F). Along with freezing temperatures these regions regularly experience whipping winds and intense seasonal changes. In the summer, days are long and bright, plankton flourish, and ecosystems are rich with food and an abundance of sea life. Come winter, the scene darkens dramatically, winds can exceed hurricane force, and ice spreads across the sea surface, causing a precipitous drop in productivity and food availability. Amazingly, life has found ways to adapt and flourish even in these, the radical and changing conditions of the Earth's high-latitudes' icehouse.

The most obvious survival strategy for living in the freezer is to have thick fur and blubber. Few animals are as well equipped in this area as the symbol of the Arctic, the polar bear. It has a heavy, two-layer fur coat, and beneath its skin, insulating fat, up to 11.5 centimeters (4.5 inches) thick. Every inch of the polar bear is covered with fur, except the very tip of its nose. In the winter, even the pads of the polar bear's feet may be encased in fur. Small ears and a nub of a tail help to reduce heat loss. Polar bears are so well insulated from the cold that when running they are at risk of over-heating. But their thick fur coat comes with a price: cleaning. Polar bears must be fastidious about staying clean and dry, because wet or matted fur offers poor insulation. To dry off after a dip in the icy water, a polar bear may shake its coat much like a dog and roll around in the snow. Some have even been seen to pick up snow in their paws and use it like a face towel. On exceptionally windy, cold, or stormy days they may dig out shelters in a snowbank and curl up inside for warmth. Pregnant polar bears dig dens in the snow or ice and may spend up to nine months inside giving birth and nursing their cubs. During those months, they remain inactive, feed-ing their young without eating, drinking, or getting rid of bodily wastes. By slowing its heartbeat a female polar bear can go into a hibernation-like state, as can the males to conserve energy when food is scarce. And both genders are big-time sleepers and like to nap, just about anywhere, anytime.

Polar bears are also well outfitted for Arctic hunting and swimming. Their white fur provides excellent camouflage for stalking prey. Actually, the hairs that make up a polar bear's fur coat are transparent and hollow, but the scattering and reflection of light makes them appear white. In cap-tivity, some polar bears have gone green when algae have colonized their hollow hair shafts; luckily a treatment of salt solution will return their color to normal. Polar bears also have no-slip bumps and small suction-cup-like cavities on their paw pads for stability and speed on the ice. They sport powerful muscles and sharp claws, along with a keen sense of smell to detect their favorite meal, a blubbery seal. Not even whales are safe from this mighty hunter. If a whale becomes trapped in the ice or shallow water, polar bears may attack and incapacitate it by cleverly biting around its blowhole. They are adept in the icy water as well, with large, webbed front paws for paddling and good underwater vision. Polar bears have been known to swim for more than 90 kilometers (55 miles) without a rest. They are true masters of the Arctic icehouse.

On the opposite end of the Earth is the charismatic penguin. Small penguins can be found as far north as the Galapagos Islands, but it is in the extremes of the frigid Antarctic where these charming and hearty seabirds prosper. There are some seventeen or maybe eighteen species of penguins worldwide, seven of which are found in and around the Antarctic. On the snow- and ice-covered land, they waddle, hop, and humorously toboggan on their bellies. In the sea, however, they are amazingly adept, swift swimmers with high maneuverability and a talent for jumping and diving. Using its wings as flippers, the penguin takes to flight within the water. It makes itself torpedo-shaped by hunching its shoulders and head and keeping its feet pressed in close to its body. Its tail serves as a rudder, steering as it flies through the sea, while it performs high-speed acrobatics, or jumps, porpoising at the surface. Penguins use their excellent underwater vision and swimming skills to hunt for their favorite foods, such as krill, squids, or fishes. Some penguin species are also relatively good deep-sea divers; emperor penguins may descend to more than 500 meters (1,640 feet) and stay underwater for more than twenty minutes. Some penguins are also expert jumpers. With a takeoff boosted by swim-power, penguins can reportedly jump three to five times their body height to land on high ice floes or rocky cliffs. In human terms that would be like a person of average height running and then high-jumping some 6.8 meters (22 feet).

Penguins are exceptionally well designed for a life on ice. They are insulated from the cold by a dense layer of smooth, overlapping feathers that have tufts of down at their shafts, which trap air and provide warmth. In the water, the trapped air protects them from the cold and provides buoyancy—essentially a custom-fit penguin drysuit. They use oil from a gland at the base of their tails to further waterproof their feathers; it adds to their insulation and helps them to glide smoothly through the water. For penguins in the Antarctic, hygiene is a matter of life or death. They must regularly preen their feathers to keep them clean, well arranged, and sufficiently oiled. To replace their worn-out feathers, most penguin species molt once a year. They typically abstain from swimming during molting and must rely solely on their blubber for insulation. Just prior to their annual feather upgrade, penguins fatten up to provide extra warmth, as they must temporarily survive the freezing cold, essentially naked.

Penguins have additional adaptations that allow them to thrive within the radical living conditions of the Antarctic. The relatively small size of the penguin's bill and flippers, and its ability to control the flow of blood to its extremities, reduces heat loss. It also has an energy-efficient, internal heat-exchange system. A penguin's arteries and veins lie close to one another, allowing blood that is heading to the heart to be warmed by blood flowing away. To reduce heat loss through bodily contact with the ice, penguins do a balancing act, tipping back on their feet to stand on their heels and tail. Their sophisticated black-and-white attire also offers benefits, both in and out of the water. On the snow and ice, their dark backs are excellent absorbers of the sun's heat, while in the water, their countershaded, two-tone dress provides camouflage from potential predators, such as killer whales and leopard seals. One species of penguin has an added and now famous means of heat conservation—huddling with friends and neighbors.

The emperor penguin is the largest of the penguins, reaching just over 1 meter (almost 4 feet) in height, and is the only species known to huddle. The remarkable story of these animals was featured in the film *March of the Penguins*. Who would have guessed a movie about seabirds in the dark, cold, and windswept Antarctic winter would boost the penguin to celebrity status and win an Oscar? The film documented an extraordinary strategy for survival and procreation in some of the most brutal conditions on the planet.

The emperor penguins' perilous story begins in about April, autumn in the Antarctic, when they arrive at a rookery along with hundreds or even thousands of other individuals, ready to breed. Scientists recently used satellite imagery to identify several previously unreported emperor penguin rookeries. Discoloration of the ice was the clue that led them to the sites, but it was not the shadow from a large number of penguins that gave them away. It was the black stains of their excrement; the penguins were tracked from space by their poo. To congregate for breeding, emperor penguins apparently look for locations with a relatively flat ice surface and if possible, some protection from the wind. Easy access to the ocean for feeding is also advantageous, but there is a catch. The rookery must be far enough inland so that as spring approaches and melting causes the ice edge to creep inland, the newborn chicks will have enough time to adequately grow and be ready to swim. It is this quandary and the timing required to ensure that there is plenty of food available when the chicks start feeding on their own that leads to the emperor penguins' grueling ordeal during the Antarctic winter.

After arriving at the rookery, the male emperor penguins court the females with elegant bows and braying. Success leads to an exceedingly brief sexual encounter, just seconds long, after which in May, the females produce a single egg that is about three times the size of a typical chicken egg. At first, the female does egg duty, keeping it warm by carefully balancing it on her feet below a fold of belly skin. Incubation may then become a team effort, with the female and male penguins carefully passing the egg back and forth, like a life-or-death game of hot potato. Should the egg be dropped and spend time on the ice, it will freeze and become unviable. The female soon hands over sole responsibility for their precious package to the male, who carefully places it on his feet, tucked under his belly. She then heads to the open ocean in search of food to replenish her own stores and to build up supplies for the impending newborn.

Because the rookery must be so far inland, female penguins may have to trek more than 100 kilometers (62 miles) to reach the open ocean and a choice meal of squids, krill, or fishes. As the females leave in search of sustenance a long procession forms, with as many as three thousand penguins marching in single file to the sea. They travel day and night, their own survival and that of their chicks at risk. Meanwhile, the male penguins remain behind to care for the eggs, without food and with little protection from the oncoming cold and wind.

Soon the worst of winter is upon the penguins; air temperatures become frigid, potentially dropping to −60°C (−76°F), while howling winds may reach up to 200 kilometers per hour (120 miles per hour). To survive these extremes, the male emperor penguins do what no others of their kind do—they huddle to share body heat. Side by side they press close together for warmth, all the while carefully balancing their precious eggs. On windy days, the males slowly shuffle from the brutal windward edge of the huddle into the protection of its lee. This way, each penguin takes a turn at the frigid outside edge of the group as well as at its toasty center. It may be more than a month before the females return, during which time the males may lose nearly half their body weight. If the chicks hatch before the females are back, the male penguins must extend their reserves even further and produce a milky substance to feed the starving newborns.

The females that make it back to the rookery arrive with their bellies full, ready to feed their offspring. To find their families, they waddle about the crowd, searching and listening for the distinctive call of their mates. Some will find their mates only to discover that, sadly, their chicks have not survived. When a female finds her mate with a healthy chick (plate 18), it is her turn to take over egg duty so that her partner can head to the sea for his first meal in months. The male and female penguins then take turns trekking to the sea for food until the chick has its waterproof feathers and is able to feed on its own. Fishing trips get less arduous as winter turns to spring, the air warms, and the ice edge approaches the rookery.

The emperor penguins' story has a happy ending if in about December or January, there is a healthy youngster frolicking in the sea and learning to forage on its own. Many, of course, don't make it. For the adult penguins, the end of their dramatic tale does not include happily-ever-after when it comes to love. They have a relatively low fidelity rate as compared to other penguin species, and most emperor penguins switch mates from one season to the next.

Polar bears and penguins are just the tip of the iceberg when it comes to organisms that are adapted to life in the freezer. From antifreeze-carrying fish to long-in-the-tooth whales, zombie sharks, and monster-sized invertebrates, an abundance of weird wildlife resides in the sea's frigid realms. During the coldest periods of the year in polar waters, when food supplies are low, some fishes can essentially put themselves on ice by lowering their metabolisms. Other fishes have special glycoproteins in their blood that act as antifreeze and quite literally prevent them from freezing. Another odd, cold-loving creature is the icefish, which has evolved the ability to obtain oxygen from seawater directly through its skin as well as its gills, taking advantage of the high oxygen content within the poles' cold water. Its blood is strangely transparent, as it has lost the need for red blood cells. The icefish also has a lightened skeleton that helps it to float and swim without expending much energy. In the intertidal zone of the Antarctic there is another organism with a different and altogether slick strategy for surviving the cold. It is a small snail, or limpet, that coats its body with a protective layer of mucus. Even in the freezer, slime comes in handy.

Some of the world's most unusual marine mammals make their homes within the oceans' icehouse. They are all blubbery creatures, but special in their own odd way. In the Antarctic there is a seal not to be messed with, known for its ferocity and its penchant for eating its relatives. The leopard seal is, like other seals and sea lions, warm-blooded, air-breathing, and flipper-footed, and must haul out of the water to rest, mate, and rear its young. Unlike the others of its kind, however, it has a distinctly spotted coat and an elongated, reptilian-like face. High on the leopard seal's list of favorite foods are other seals, along with penguins, squids, and fishes. When these salty dishes are unavailable, it will also eat krill. The leopard seal's teeth are well designed for a diverse bill of fare. The canines and incisors are built for gripping and tearing prey, while its interlocking molars make a perfect sieve for feeding on zooplankton.

Found on the opposite side of the world, within the Arctic, is another unusual relative of the seal and sea lion, the walrus. Its rotund shape, cinnamon-brown coloring, whiskered, flat face, and large tusks make it conspicuous among its peers. Walrus tusks are actually overgrown canine teeth, and are most impressive in the males, reaching up to 100 centimeters (over 3 feet) long. Walruses use their tusks like ice axes to pull themselves up onto the ice or land. They also come in handy for fighting and for foraging for clams, crustaceans, and other invertebrates at the seafloor. To search for and collect food the walrus may sled through the sediments feeling around with its very tactile and maneuverable whiskers. A strong wave of the foreflipper may help to sweep the seafloor, or the walrus may use its strong cheek muscles to create a powerful jet of water to blast away the bottom sediments. Once discovered, their favorite food, the clam, has little defense against this supersucker—the walrus can literally suck the organism right out of its shell and then swallow it whole. They are the shucking champions of the sea, reportedly eating as many as sixty clams in just one twenty-four-minute dive! Walruses are gregarious animals and tend to sleep close together in herds when hauled out. Most of the year, except during breeding, the sexes stay somewhat segregated, with the males tending to prefer pebbly beaches, while the females like to haul out on drift or pack ice.

Also within the Arctic are two closely related and somewhat secretive whales, the narwhal and beluga. The mottled gray narwhal has a streamlined body up to about 5 meters (16 feet) long, a slight hump instead of a dorsal fin, and one very bizarre, overgrown tooth. Its long spiraling tooth, or tusk, juts out from the left side of its upper jaw and is believed to have inspired legends of a more mythical creature, the unicorn. The true purpose of its tusk has been a matter of conjecture for centuries. The females rarely have tusks, and the males have been seen to cross theirs as if in swordplay, leading some scientists to suggest that they may be used in dominance displays or are suggestive of virility. Other researchers propose that the tusk is used in feeding, to break through the ice, or that it may have environmental sensing capabilities. The mysterious narwhal is among the oceans' deep-sea divers, regularly submerging to depths of over 1,000 meters (3,280 feet), presumably to feed on squids and fishes such as the halibut.

And as for belugas, let me just be up front about my bias in this matter: I am completely enamored by these enigmatic, endearing, and altogether fascinating creatures. During a behind-the-scenes tour at the Georgia Aquarium I walked by the top of the beluga tank. Two creamy white whales abruptly stopped their rolling swims to bob upright and curiously look my way—a stranger among the aquarium's staff. As I passed by, their amazingly flexible necks allowed their heads to swivel and their eyes to follow me as they intensely watched my every move. They had me at hello.

The beluga whale is a streamlined, blubbery resident of the Arctic's extremes, whose unusual white coloration allows for camouflage within the ice-covered seas. They can make long swims under the frozen surface, are one of the oceans' most vocal residents, and may have the most sophisticated navigation system on the planet. The belugas lack a classic dorsal fin, but instead have a long ridge running down their backs, which is used to break through the ice and create breathing holes. One of the beluga's most astonishing and unusual features is its extremely flexible neck due to unfused cervical vertebrae. The supple blubber about its head and short beak also enables it to change the shape of its mouth and melon, the fat-filled, bulging part of the beluga's forehead just in front of its blowhole. This allows for some of the whales' most endearing and curious facial expressions as well as their observed talents for blowing bubble rings and blast-spitting, like the walrus. The belugas use echolocation much like other whales to navigate and detect prey such as fishes, squids, crabs, and snails. Unlike other whales, however, the beluga is able to change the shape of its melon to focus its echolocating beam of sound more precisely. Belugas may even be able to "see" around corners by bouncing sound off ice floes. The beluga is sometimes called the "sea canary" due its remarkable ability and propensity to produce sound. While individual animals have distinct calls, in general they are able to create the most diverse array of sounds of any marine mammal, with a repertoire that includes whistles, squeaks, trumpeting, roars, pulsed calls, and much, much more. They are the male humpbacks' stiffest competition in the ocean idol competition.

In the summer, belugas crowd into shallow coastal areas to give birth and to molt, the latter an unusual phenomenon for whales. By rubbing their bodies on the bottom they remove aged yellowy, wrinkled skin, which is replaced by smooth white growth. Females give birth every two to three years and have a close association with their calves, which start life wrinkly and gray, nursing for up to two years. The belugas' travel in the fall and winter, but exactly where they go remains a mystery.

Like some other whales and dolphins, belugas will come to the aid of sick or injured comrades, even putting themselves at risk. Both the beluga and narwhal must be wary of polar bears and killer whales, though the greatest danger to these and other marine mammals has been and continues to be humans. The white whale is a unique, mysterious, and engaging creature that is well adapted to the oceans' radical ice-cold conditions. And to answer a question that is often asked about belugas at aquariums (really)—no, there is no connection to the similarly named caviar, which is the roe or eggs of a fish, the beluga sturgeon.

Within the extreme cold of the northern high latitudes there is another strange and elusive creature. It has been described as spooky and ghoulish, and the search for it a bit like chasing after bigfoot. The stout, small-eyed Greenland shark is one of the oceans' largest but least known predators, reaching lengths of some 7 meters (23 feet). It is darkly colored with a small, flattened triangular dorsal fin and a large asymmetrical tail (plate 19). A lover of the cold, it stays deep in the sea during the summer, but may swim up to the very edge of the ice in the winter. Greenland sharks are believed to grow extremely slowly and live long, possibly up to two hundred years. They are formidable predators, easily able to take a chunk out of a seal or narwhal or suck up a huge halibut from the seafloor, like a giant vacuum cleaner. They will devour the living or consume the remains of the dead. Within the stomachs of the few Greenland sharks that have been caught, a gruesome diversity of remains have been found, including the remnants of polar bears, porpoise, whales, birds, other sharks, and bottom-dwelling invertebrates, along with a dog, horse, and even a reindeer. The Greenland shark's jaws are well outfitted for dining on the sea's denizens, with dagger-like upper teeth to hold prey in place and serrated lower teeth to do the slicing. Because this odd, reclusive shark lives in icy-cold water, scientists assume that it has a low metabolic rate and must therefore be rather sluggish. But then how does it successfully catch such a wide variety of prey? Is it really a slow-moving zombie, or is there something more to this mysterious creature that we have yet to discover? Some researchers speculate that a parasitic copepod regularly found embedded within the shark's eyes makes it blind, but also acts as a dangling, bioluminescent lure for potential prey. Then again, maybe it is just that the Greenland shark devours anything and everything that it can get its very large mouth around or suck in.

Scientists are discovering that the oceans' ice-covered realms are, in fact, teaming with life. Even within the ice itself there is growth, as algae colonize channels of brine. Beneath the frozen surface, in the water and on the seafloor, there is a virtual menagerie of cold-loving creatures, and many of these organisms are surprisingly large, even gigantic. There are sea stars over 0.5 meters (2 feet) wide, sea spiders with legs that are 50 centimeters (1.5 feet) long, and towering sponges that grow up to 1.5 meters tall (about 5 feet). Jellyfish of nightmarish proportion have also been found, with enormous swimming bells and trailing tentacles some 8 meters (26 feet) long. It is unclear what causes this polar gigantism, but scientists theorize that it may be related to the high oxygen content of the water, the cold temperatures, or the organisms' decreased metabolic rates. We don't know very much about many of the organisms that are found within the oceans' freezers, but so far, they are proving to be fascinating creatures that are surprisingly well adapted to living in radically cold conditions.

HOT AND TOXIC

For most organisms on the planet, proximity to a superheated, metal-rich plume of acidic seawater in the deep sea would spell certain death. But what is fatal for some gives life to others. The abundance of organisms found at deep-sea hydrothermal vents shocked the world when it was first discovered some thirty years ago. Deep-sea vents are now known to occur throughout the world's oceans and to host communities extraordinarily rich with life. More than five hundred species have been described from these unusual oases at depth, and many of the creatures found had never before been seen. Among the most famous of vent creatures is the giant tubeworm, reaching 1.5 meters (5 feet) in height and topped by a blood-red, feathery, gill-like plume. There are also enormous clams and mussels, hairy snails, furry worms, colorful crabs, and a colonial animal that bares a striking resemblance to an ethereal pinkish-orange dandelion tethered to the seafloor like a hot-air balloon.

Hydrothermal vents occur in the deep sea at volcanically active sites along the Earth's midocean ridges, on seamounts, and in submerged basins adjacent to volcanic islands. At these locations, seawater percolates down through cracks in the seafloor, is heated by underlying magma, and then periodically erupts through fractures in the seabed. As the erupting fluid mixes with the surrounding bottom water, minerals rapidly precipitate out, creating smoke-like plumes and towering chimney-shaped deposits on the seafloor. Dark and deep and under tremendous pressure, it is already an extremely hostile setting: then add seawater blasting out of the bottom that is superheated up to about 350°C (662°F), corrosive, and concentrated with heavy metals, which are toxic to most animals. Amid all of this, however, there can be a flurry of life in the form of snow-like particles swirling about the seafloor, creating a deep-sea whiteout or covering the bottom in a soft, pallid blanket. These are the microbes that are the foundation of the vent food web and ultimately create the strange communities that we now know can survive far from the life-giving energy of the sun.

In the sun-driven food web, organisms use sunlight to drive photosynthesis to produce organic carbon from inorganic materials. In contrast, at hydrothermal vents microbes use chemicals, such as hydrogen sulfide, to fuel growth in a process known as chemosynthesis. These chemosynthetic microbes (bacteria and a newly identified form of life named Archaea) are the food for the larger organisms of the vent community or are part of a symbiosis that enables them to survive.

The celebrity species of the vent community is undoubtedly the aforementioned red-plumed giant tubeworms (plate 20). They are gutless, mouthless, and buttless wonders of the deep sea that can grow in dense thickets about active vents. Each worm is sheathed in a cream-colored, banded tube made of chitin (like the exoskeleton of a crustacean) and protein. The coloring of the worm's red plume comes from hemoglobin in its blood, which is used to transport sulfide, oxygen, and carbon dioxide from its plume to a large cavity lower in its body. Bacteria living within this cavity use the transported chemicals in chemosynthesis to grow, providing nourishment for their worm host. Tubeworms, like other vent creatures, cannot live directly within the hottest hydrothermal fluids or they would literally be cooked. The Pompeii tubeworm, however, can really take the heat and may sit with its rear end in scalding hot water, up to 80°C (176°F). Its body appears oddly covered in white or bluish hair. This fuzzy facade is due to the growth of filamentous bacteria, which produce a heat-resistant enzyme and enable the worm to bathe within a really hot, hot tub. The Pompeii worm can grow up to about 15 centimeters (6 inches) in length, and appears to both feed on and cultivate bacteria.

Within the different ocean basins, the makeup of vent communities varies. In the Pacific, hydrothermal sites are dominated by tubeworms, clams, and mussels, whereas in the Atlantic the name of the game is mussels and shrimp, lots of them. Incredibly dense swarms of shrimp, with as many as 2,500 individuals per square meter, have been found at Atlantic vents, aggregating where the temperature is about 10°–30°C (50°–86°F). To avoid accidentally becoming shrimp-on-the-barbie, these animals are outfitted with their own undersea heat detectors. They are able to avoid the barbeque using a sort of modified eye on their backs that can sense the very dim light emanating from the thermal glow of the hottest vent waters. The shrimp feed principally on bacteria, which they collect, cultivate, and consume. In the Indian Ocean, vents appear to harbor communities much like those found in the Pacific, except that they may also have an abundance of Atlantic-type shrimp.

When discovered in 2005, one unassuming member of the deep-sea vent community became somewhat of a celebrity, with its photo plastered across the Internet and in the pages of magazines. It is a seemingly simply creature, a small whitish crab with a little somewhat flattened, egg-shaped body, pointy head, and two very long front claws that appear covered in thick yellowish fur. Filamentous bacteria make up its fur coat, which earned it the popular name "the yeti crab."

Also among the residents of the vents are clams the size of dinner plates that host symbiotic bacteria and cluster within fractures at the seafloor. Joining the large hydrogen-sulfur-laced clams are giant mussels, along with hairy snails, bristle worms, and other crabs. Brittle stars crawl about, wedged in among the clams or hidden amid the tubeworms. There are also large, stalked barnacles that look a bit like miniature palm trees, slender, worm-like sea cucumbers, a host of other snail and shrimp species, and an unusual colonial creature that mystified scientists when it was first discovered. It has a pinkish-orange ball at its center that is surrounded by small feeding polyps and has been observed swimming or floating just above the seafloor attached by thin, stringy tentacles—tethered like a huge balloon at the Macy's Thanksgiving Day parade. The delicate dandelion creature is a siphonophore, or colony of polyps, like the Portuguese man-o-war, but we know little more than that.

Animal life often extends out from the active hydrothermal vents and may include rocky fields of flower-like sea anemones and stalked jellyfish. Larger predators lurk about as well, including several deep-sea octopus species that have proven difficult to capture. One vent-visiting octopus is warty and pink, while another appears ghostly white with relatively long and slender arms. There are eels, as well as fish that look like eels, and a peculiar shark that sluggishly swims about in search of prey. It is the pallid ghost shark, a creamy to light grayish fish that apparently grows up to

PLATE 11.
A fish stops in
for a clean and
shine. Photo by
Steven Kovacs.

PLATE 12. The elegant beauty of a pacific sea nettle. Photo © David Wrobel/ SeaPics.com.

PLATE 13. A male octopus passes his seed to a female with his specialized baby-maker arm. Photo © Steven Kovacs/SeaPics.com.

PLATE 14. Squid sex and eggs off California. Photo © Mark Conlin/ SeaPics.com.

PLATE 15. A pygmy seahorse in camouflage on a sea fan, Borneo, Malaysia. Photo by Vickie Coker.

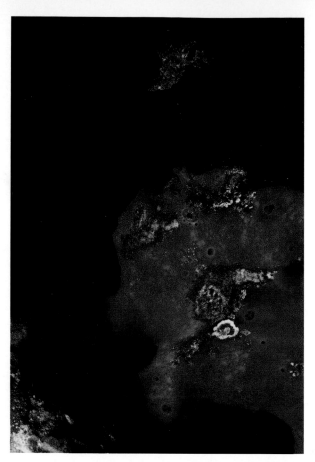

PLATE 16. A Pacific frogfish angling for a meal. Photo © D. R. Schrichte/SeaPics.com.

PLATE 17. Humpback whales bubble-net feeding in Southeast Alaska. Photo © Michael S. Nolan/SeaPics.com.

PLATE 18. A pair of emperor penguins and their chick as the brutal Antarctic winter begins to turn to spring. Photo by Fritz Poelking/SeaPics.com.

PLATE 19. The elusive Greenland shark, St. Lawrence Estuary, Canada. Photo ©
Doug Perrine/SeaPics.com.

PLATE 20. Lush deep-sea vent community starring *Riftia pachyptila* tubeworms at
a depth of 2,500 meters along the East Pacific Rise. Photo courtesy of Richard Lutz,
The Stephen Low Company, and the Woods Hole Oceanographic Institution.

PLATE 21. (A) Bone-eating zombie worm female dissected from a whalebone, and (B) whalebone and worms collected from 600 meters in Monterey Canyon, California. Photos by Greg Rouse, Scripps Institution of Oceanography.

PLATE 22. A huge mound of conch shells laid to waste in Bonaire. Photo © Amar & Isabelle Guillen/SeaPics.com.

PLATE 23. Trash washed up on Kanapou Beach, Island of Kahoolawe, Hawaii. Photo provided by NOAA Marine Debris Program.

about 1.3 meters (4.3 feet) long. It has a large head and trunk tapering to a pointed, eelish tail along with big eyes and a snout that tips up almost snobbishly. The pallid ghost shark also has an odd, conical mouth and relatively large pelvic fins, and running along its body and across its head are distinct black lines that seem to divide the fish into segments like a jigsaw puzzle. These are its lateral lines, which sense vibration and movement in the water. The pallid ghost shark feeds on the small fishes and invertebrates that dwell within this surprisingly lush hotbed of life.

The variety and abundance of organisms living at deep-sea vents has inspired imaginative location names such as the Garden of Eden, Magic Mountain, Lucky Strike, and Zoo Chimney Field. Within vent fields, researchers often identify specific sites based on the communities found, such as Clam Acres, Fish Spa, Barnacle Beach, Bacterial Balls, and Mussel Slope. Other names like Inferno and Brimstone Pit reflect the hellish conditions that occur.

Dead tubeworms and clams found at inactive vents are evidence that these communities are relatively ephemeral features linked to periodic volcanic activity. The comings and goings of such communities are facilitated by the fact that many vent animals have planktonic larvae, which are believed to play an important role in how organisms spread locally and how they colonize newly active vents that may be hundreds of kilometers away. Rising warm-water plumes at active vents appear to entrain larvae and carry them into currents, which may be constrained by the bottom topography. If retained within circular eddies, the larvae may remain in the area; otherwise they may be transported farther away, possibly traveling for weeks in search of the heat or chemical cues that signal an active vent site appropriate for settlement. There is still much to learn about the strange creatures that thrive in the deep sea's radical hot and toxic conditions.

COLD AND TOXIC

Communities that are similar to those found at deep-sea hydrothermal vents have also been discovered at cold, hydrocarbon seeps in the Gulf of Mexico and Mediterranean. The residents of these deep-ocean oases include giant, bacteria-hosting clams, tubeworms up to about 3 meters (10 feet) long, along with mussels that harbor microbes living off of methane as well as snails, shrimp, and fishes. In association with the Gulf of Mexico's methane seeps, scientists have also found curious seafloor pools and lakes of superdense, supersalty seawater. They are created as salt from underlying mineral deposits leaks upward and mixes with seawater to form brines that are four to five times as salty as typical ocean water. This brine settles in depressions in the seabed to create eerie deep-sea puddles, pools, and even large lakes, which are sometimes lined by mussel-rich seep communities. But animals living at the edge of these briny deep-sea pools must take care, because falling into their depths means truly getting into a pickle, and a salty death.

Also in the deep Gulf of Mexico are deposits of methane hydrate, an icy crystallized form of methane gas. Researchers were surprised to discover another unexpected and odd worm living within the methane hydrates. Dubbed the "ice worm," it creates and resides within icy burrows and feeds on chemosynthetic bacteria.

DEEP, DECAYING, AND UNDEAD

In 2002, within the deep waters of Monterey Bay, California, scientists discovered another type of worm, maybe the most bizarre yet. Distant relatives of the celebrated vent tubeworms, these slender, slimy creatures have a macabre obsession—they colonize and feed on the bones of dead whales that have sunk to the seafloor, or as scientists put it, at "whale falls." The worms live in gelatinous tubes and are typically several centimeters long and about pencil-wide. They are topped with feathery red plumes somewhat like their relatives, but at their base they have fleshy, bile-green roots, jam-packed with bacteria (plate 21). As they tunnel into a dead whale's bones, the bacteria in the worms' roots digest the fat and oil within, thereby obtaining nourishment for themselves and their worm hosts. It is a zombie worm feast that surprisingly only one gender partakes of. While the female worms chow down on the dead, the males (once again) are relegated to a life as microscopic sperm producers, this time living in harems—inside the females. The dwarf males are fed by the lipid leftovers of the eggs they developed from and grow little past the larval stage. They produce sperm that move up through the female tubeworm and fertilize her eggs, which are released. The harem within a large female may contain hundreds of males, which collect within her over time. Scientists think that in an unusual twist of fate, the sex of the larvae is determined by where they settle. Larvae that settle on whalebones turn into females and grow. Larvae that land on female worms, however, turn into males and are destined to join a harem in the ultimate of matriarchal societies. Bone-eating worms have also been discovered in other areas of the Pacific Ocean and in England's North Sea. The worms in the North Sea were found in shallower depths (120 meters or 394 feet) on a dead minke whale and are a different species. When these worms were disturbed, they retreated into their gelatinous tubes and released gobs of mucus, hence the oceans' own "bone-eating zombie snot worms." No horror movie needed here—there really are zombies that devour the dead on planet Earth—slimy, bone-eating ones in the deep sea.

WHY THEY MATTER

Like all the other creatures described so far, each and every organism portrayed in this chapter plays an important role in the oceans, from the snow-like microbes that are at the base of the deep-sea hydrothermal vent food web to the predacious polar bear, the Arctic's top hunter. They also have both familiar as well as surprising connections to humankind and society.

In the Arctic region, native populations have long hunted whales for subsistence purposes and revered them as a part of tradition and culture. Seals and walruses have also been taken for food and for their pelts, and in the past the walrus was killed for the ivory of its tusks. Such practices, though now limited, have become the focus of great controversy as people increasingly decry the killing of marine mammals for food, clothing, or tradition.

Alive and on view, several animals from polar climes are now some of the oceans' greatest ambassadors. At aquariums and zoos, beluga whales, polar bears, and penguins attract huge crowds, while seals, sea lions, and walruses also make for popular attractions. There is an argument to be made against keeping these and other marine organisms in captivity; however, there is also a strong case for showcasing them *if* they are kept in good health and in humane conditions, are treated well, and are either rescued or bred in captivity to negate wild capture. These animals may be our best hope to educate people, build an emotional connection to the sea, and inspire lifelong caring and stewardship. The reality is that most people will never see the oceans' animals in the wild firsthand, especially those that reside within the most extreme environments. Zoos and aquariums also provide an opportunity to study marine organisms and rehabilitate those that are hurt or sick, and most now have important conservation and breeding programs.

Human Health and Biotechnology

Polar bears and human health: Is there a connection? Clearly, a polar bear can be exceptionally detrimental to human health when a person comes face-to-face with this massive and extremely dangerous predator. On the other hand, the polar bear's biology may prove surprisingly beneficial to the health of millions of people. In traditional Chinese medicine, powdered bear bile has historically been used to treat diseases of the liver and gallbladder. In 1901, Swedish researcher Olof Hammarsten identified an active compound, now known as ursodeoxycholic acid, in the bile of polar bears. It was later found to play a role in how they metabolize their high-fat diet and store adequate reserves for denning without detrimental side effects. Ursodeoxycholic acid is now chemically synthesized and used to treat human diseases and ailments that involve bile, such as cirrhosis of the liver. Biomedical researchers are also studying how pregnant polar bears den, without ill effects, for many months. Humans who are inactive for similar lengths of time experience significant muscle and bone loss. During denning, it appears that polar bears are somehow able to recycle calcium lost from their bones back into new bone growth. Understanding how this works could have significant application to a widespread human disease, osteoporosis. The polar bear's ability to reabsorb its urine back into its bloodstream while denning is also being investigated with regard to renal disease and kidney failure. In addition, polar bears are being studied to better understand and treat human obesity and diabetes.

The strange icefish of the poles, with its transparent blood and reduced skeleton, is also of interest to biomedical researchers. Their lack of red blood cells is being studied to learn more about how blood cells are formed and regulated, and the reduced mineralization of their skeleton may provide additional clues to combating osteoporosis. Researchers are also investigating the growth of blood vessels in icefish and how it might help to develop a means to shut off blood supply to cancer cells.

At the oceans' deep-sea hydrothermal vents and cold seeps, it is the microbes that seem to hold the greatest promise in biomedical research and biotechnology. An enzyme from bacteria living in the hot springs of Yellowstone National Park has already proved a highly valuable commodity for use in the study of genetic diseases, in forensic analysis at crime scenes, and in genome sequencing. Researchers are particularly interested in the enzymes that enable deep-sea vent bacteria to withstand high heat or use substances such as methane or hydrogen sulfide, which are usually toxic. The heat-resistant enzymes from the Pompeii worm's bacteria could prove useful in drug production and food processing. And the next-generation superdetergent could come from the bone-eating zombie worms and the enzymes their bacteria use to digest fat and oil. Organisms living at deep-sea vents and seeps are also being investigated for use in the cleanup of toxic wastes, in the production of biofuels, in tissue regeneration, and in the fight against cardiovascular disease. Using DNA and molecular technology, scientists have already identified tens of thousands of different microbes living in the deep sea in extreme conditions. It is an area that continues to be ripe for discovery and rich with promise.

Life in the oceans has evolved over billions of years, and organisms now live in a wide range of conditions, including some of the planet's most radical. Some scientists believe that deep-sea hydrothermal vents were, in fact, where life first began. Other researchers are looking to the extremes of the oceans as potential analogs for distant planets and alien life forms. As we explore more of the world's oceans, we will continue to discover new and strange forms of life, living even in unexpected places. These creatures, no matter where they live or how bizarre they may be, are all part of the oceans' vast web of life. It is this wondrous and great diversity that is the foundation of the sea's productivity and that allows humankind to depend on a wealth of marine resources. It is also this great diversity that is now at risk.

9 Danger Looms

Danger is ever-present in the sea; it typically comes with sharp teeth, sticky slime, and lethal venom. But for years there has been a growing menace in the oceans, a deadly predator whose diverse and long arms are global in their reach. The victims of this serial killer are numerous; the loss of life is disturbing, and the potential consequences for the oceans and humankind catastrophic. Who is this killer and the cause of such destruction? Unfortunately, the answer is us. Human activities have dramatically impacted the sea's populace, destroyed critical habitats, and caused a serious immigration problem in the oceans. From the mega-slime-producing hagfish to the deadly cone snail and clever beluga, few, if any, residents of the sea can escape the toll that has been wrought by humankind. It threatens the very diversity that is so critical to the oceans and to the precious benefits that we obtain from them.

THE DECLINE OF RESIDENTIAL LIFE

Humans have literally changed the demographics of the oceans. Up to 90 percent of the sea's large predatory fish are reportedly gone. In places where cod, salmon, oysters, or the small menhaden used to exist in vast numbers, far fewer are present today. Regional stocks of fish across the world have declined, and most are overexploited. The bounty of abalone in the Pacific Northwest, spiny lobster in the Caribbean, and queen conch in Florida and elsewhere is all but gone (plate 22). In Europe, eel populations have decreased drastically since the 1970s, and only a few percent of these once abundant creatures are now mysteriously making their way into the Sargasso Sea. Tens of millions of sharks are stripped from the sea each year and killed, many just for their fins. Shark expert Neil Hammerschlag fears for the survival of the fascinating hammerhead because their fins are the most prized in the trade and their populations have plummeted by some 90 percent in the western Atlantic in just the past couple of decades. We pull seahorses, sea cucumbers, and the mighty bluefin tuna and sword-fish from the oceans' breast in enormous quantities, and some countries continue to allow whaling under the guise of research. Overfishing and unsustainable practices are reducing the populace of the sea and literally killing the lifeblood of the oceans.

Humankind has long depended on fish for food, jobs, tradition, and recreation. The oceans have essentially been considered a common good, and many people believe it is their right to fish where and however much they want. But over time, as more fish are caught with ever-improving technology, there are consequences to this way of thinking, and overfish-ing is a big one. Some species are long-lived, slow to mature, or have a low reproductive rate; these varieties, such as orange roughy or Chilean sea bass, are especially vulnerable to overfishing. Consider the sand ti-ger shark, whose pups get an early in-womb education in hunting. These sharks have only one or two pups a year and may not even give birth on an annual basis. If more individuals are annually removed from the sea than are added by reproduction, the sand tiger shark and other such fishes may go the way of the dodo.

The detrimental impacts of overfishing can also move down and across the food web. When top predators are overfished, their prey species may multiply and change the dynamics of an ecosystem. Off the east coast of the United States, large sharks have been overfished, leaving the popula-tion of one of their main prey species, cownose rays, unchecked. In 2004, due to a proliferation of scallop-feasting cownose rays, a century-old bay scallop industry in North Carolina was forced to shut down. As we remove the largest fishes from the sea, some species are also getting smaller; they are just not growing as big as they used to. Because larger fish of-ten produce more eggs, removing the big fish also reduces the number of potential offspring, thereby significantly impacting the ability of popula-tions to sustain their numbers or to recover from overfishing. In addi-tion, as traditionally caught species disappear, others types of fishes are being exploited, and again, the impacts can transfer down and across the food web.

The bottom line is that without a sufficient number of individuals, populations in the oceans cannot maintain proportions large enough to sustain our use, or in some cases, even to breed successfully. Think about the orgy-loving queen conch. Reproduction in conch depends on having enough players at the party—it is density-dependent. Below a certain number queen conch will not breed, and they will eventually head toward extinction. In some areas, fishermen have targeted places where fish are known to aggregate, leading to a rapid and rich catch. These aggregations, however, are often for spawning, aka group sex, meaning that fishing them is a surefire and quick way to catastrophe. Loss of individuals due to overharvesting, sickness, injury, chemical disruption of reproductive processes or biology can all render marine populations at risk. When you cannot find a mate, you cannot have sex, and then, well—no babies. And if your baby-making machinery is not functioning properly—no offspring there either. One wonders why it has taken us so long to recognize something so simple.

For some species, aquaculture may offer an alternative, but the environmental benefits of fish farming can be negated when mangroves or other coastal habitats are destroyed to build facilities, wild populations are caught as brood stock or used as feed, or concentrated wastes are released into nearby waters. Other issues that must be dealt with include preventing the escape of farmed fish so that they don't spread disease or reduce the genetic strength of natural populations, and the overuse of chemicals and antibiotics.

WASTEFUL PRACTICES

The residents of the sea have it easy when it comes to wastes; their discards are either eaten by other organisms or recycled within the marine ecosystem through processes such as decomposition and reuse. If only human wastes worked that way, from the excrement and chemicals we flush down the toilet to the plastic debris we throw into the trashcan. Unfortunately, our wastes are much harder to deal with and we have yet to find ways to efficiently recycle or reuse most of our discards (plate 23). In some areas of the world, untreated and poorly treated sewage is still being released into the oceans, making them unhealthy for bathing or swimming and at times fueling the profuse growth of algae. Excess fertilizer from agricultural lands can spur harmful algal blooms and contribute to the creation of dead zones. Power plants and other industries emit aerosols (and mercury) that spread throughout the atmosphere and fall into or are absorbed by the oceans. Polluted water running off the land after storms and in rivers and creeks brings unwanted garbage and contaminants into the oceans.

With the exception of a few, notably large oil spills, most of the oil and plastic entering the sea comes from the land, not from ships or accidents as many people think. Over the years, currents have carried our plastics far from land and from sight, converging in areas such as the "Great Pacific Garbage Patch," where our unsightly debris is creating the world's largest rubbish dump. And those convenient plastic bags that are just about everywhere—when floating in the sea they resemble jellyfish, tempting sea turtles and other creatures to partake of what may be their last supper.

It is the citizenry of the sea that suffers most from our wasteful practices. Marine animals accumulate toxins in their bodies, and ingest or strangle in our plastic debris and oil. As we flush away our wastes, with them also go the drugs and chemicals we have ingested or dumped, from birth control pills to caffeine. It is unclear what impacts these contaminants are having on marine life, but the initial findings are worrisome.

Chemicals have also been found to move up through the food web and accumulate in marine animals over time. In the Pacific Northwest, killer whales have been called the most PCB-contaminated animals on the planet. In addition to PCBs (polychlorinated biphenyls), the whales have high levels of dioxin and chlorinated pesticides in their bodies, all of which appear to come principally from one of their main food sources, chinook salmon. The salmon pick up contaminants by living in waters polluted by fallout from the atmosphere, agriculture, pulp mills, military bases, and urban runoff. High toxin levels may suppress the killer whales' immune systems and contribute to the decline of their populations in the region. Not even the champion of the predator pentathlon can defeat this insidious killer. Failing salmon stocks may also be a factor. Narwhals and the charismatic belugas, along with many other marine mammals, are at risk due to contamination and population declines.

In fishing, wasteful practices are compounding the problems posed by our ever-growing demand for seafood. Longlines used to catch fishes, such as swordfish and tuna, regularly hook creatures that are too small to legally keep as well as hundreds of thousands of sharks, dolphins, sea turtles, and sea birds, all of whom are unceremoniously tossed back into the oceans dead or dying. At least seven million tons of this "bycatch" is annually discarded as waste. Bottom trawlers that scrape the seafloor for shrimp, scallops, or bottom-dwelling fishes are responsible for much of the worlds' bycatch and are literally bulldozing the neighborhoods of the sea, like clear-cutting a forest for its deer. Industrial-sized fishing operations cause more destruction and waste than smaller-scale efforts, except in areas of the world where people use dynamite or bleach on coral reefs to "harvest" their catch, destroying the very habitat upon which their livelihoods are built. Our wasteful practices threaten marine life and the resources we have so grown to depend on.

GLOBAL CHANGE

The citizens of the oceans have long existed in a changing world. They now, however, face a shift in climate that scientific evidence suggests is unnaturally rapid. Throughout its history, the Earth has gone through cycles of warming and cooling, driven by a variety of factors, such as the amount of solar radiation hitting the planet, the quantity and distribution of ice on the surface, and the concentration of specific gases in the atmosphere. Since the industrial revolution, the levels of carbon dioxide and other "greenhouse" gases in the atmosphere have measurably increased. In conjunction, our climate has warmed at an accelerated pace, faster than what scientific evidence suggests is natural, and with it, sea level and seawater temperatures are rising, the oceans are becoming more acidic, and there is potential for more severe weather, flooding, and coastal erosion. Already, organisms in the oceans are feeling the heat. In response to the warming climate and oceans, some species are shifting their range, and the abundance of others is changing. Recent research suggests that phytoplankton populations are declining, kelp forests are becoming less widespread, and in the Antarctic, krill are being replaced by salps, while some ice-dependent species of penguins are fast disappearing. Some marine life may not be able to adapt fast enough or shift their distribution in time. Corals may be the poster children for the problem; worldwide they are exhibiting increased episodes of bleaching, disease, and mortality. In the Arctic, the signs are also clear: polar bears are searching farther and farther for ice to rest on and to feed from, and in the summer the Earth's most northern ocean may soon be a relatively ice-free zone.

Temperature in the ocean also exerts a strong influence on fundamental biologic processes, such as metabolism, growth, and reproduction; all of which may be altered as the thermometer rises. For organisms whose sex is determined by temperature, warming may change the gender makeup of populations. For example, the sex of baby sea turtles is dependent on the temperature of the nest in which the eggs are buried. If the nests become universally too warm, we may soon have a female-dominated sea turtle population without enough males to bring forth and ensure future generations. The increasing occurrence and frequency of dead zones, jellyfish blooms, and algal outbreaks may also be caused in part by climate change. We cannot predict exactly what the impacts of climate change are going to be in the oceans (or throughout the world), but two things are clear—they are already happening, and it is just the beginning.

HOUSING LOSS

When habitats are lost in the oceans or along our coasts, the residential living space for the sea's creatures is reduced. The loss of mangroves, seagrass beds, and wetlands due to development, dams, or pollution destroys the oceans' nurseries, where the young of many marine species spend their formative years. The loss of tropical beaches means fewer nesting sites for sea turtles. Bottom trawling wipes out entire seafloor neighborhoods, and as coral reefs and kelp beds disappear, so too does the housing for thousands, if not millions, of organisms. In 2008, experts estimated that nearly 20 percent of the world's existing coral reefs had already been lost, and that another 35 percent were at risk. The outlook is even bleaker today. The disappearance of ice in the Arctic threatens polar bears, seals, and the other animals that rely on it for food, a place to breed, and protection. Changes in the Antarctic spell a warning for the charismatic penguin and other inhabitants of the southern icehouse. And as noted earlier, as the chemistry of the oceans changes, the creatures that build their homes, skeletons, or shells from seawater are especially at risk. Included among the threatened are foraminifera, pteropods, the iconic queen conch and the lobster, cone snails, corals, and more. Housing loss is a serious issue in the oceans.

AN IMMIGRATION PROBLEM

The control of border crossings and immigration is not just a problem on land; it poses a serious danger in the oceans and is well illustrated by an army of foreign invaders taking up residence in the Bahamas. Lionfish naturally live and breed in the Pacific; however, by 2009, thousands of lionfish were found living and reproducing in the Bahamas. Without a natural predator, they are now gobbling up fishes, crustaceans, worms, and whatever else they want from the region's coral reefs, and breeding without restraint. Lad Akins is the special projects director for the Reef Environmental Education Foundation (REEF) and has been spearheading their lionfish research and eradication program in the Bahamas. His team has found over fifty different species of prey in the stomachs of lionfish, including both commercially and ecologically important species, at sizes up to half that of the lionfish themselves. Akins and other experts fear that these invaders will have a catastrophic impact on the region's coral reefs and are spreading uncontrollably. Lionfish have already been spotted in many other areas of the Atlantic and Caribbean, including Florida, North Carolina, the Cayman Islands, and Belize. How did this problem start? Pacific lionfish are popular in aquariums, but they can grow quite large and can be aggressive. It is believed that the initial invaders were castoffs from the aquarium trade or their owners, and that at least a few were released from a seaside aquarium in Florida that was damaged during Hurricane Andrew in 1992. Complete eradication or control will be difficult, and the Pacific lionfish is now well on its way to establishing residency in the waters of the Atlantic and Caribbean.

The lionfish is not the only marine organism that has shown up in places where it does not belong. Across the world, hundreds of non-native species have crossed borders and immigrated into waters they are not naturally found in. In San Francisco Bay alone, there are over two hundred non-native species now living within the region's waters, residing within its sediments, or attached to its rocks. An army of zebra mussels has invaded the Great Lakes, and Asian and European shore crabs are now living among the natives of California and New England. In Chesapeake Bay, a foreign pathogen is believed to be responsible in part for the dramatic decline of the local oyster population. And throughout the world, non-native jellyfish are spreading and sliming their way into local waters. The problem of invasive species is widespread and growing fast.

Non-native species are a problem because they can outcompete local organisms, cause damage to fisheries, foul infrastructure, and alter ecosystems. The impacts of these invaders are also costing us billions of dollars. It is estimated that from 1989 to 2000, the economic impact of zebra mussels in the United States may have been as much as a billion dollars. In California and other areas, millions of dollars have been spent to control the spread of an invasive green alga, *Caulerpa taxifolia*, while in the Great Lakes enormous amounts are being spent to prevent the invasion of Asian carp.

Many marine immigrants make their way into foreign waters as stowaways in the ballast water and holding tanks of large ships. It is estimated that every day more than seven thousand different species of marine life travel around the world in and on ships; additional transport is provided by the aquarium trade, fishing and recreational boats, and floating marine debris.

YOUR FIRED!

Who is responsible for the woes of the sea's populace or our reduced benefits and increased costs due to a decline in the oceans' health? In some ways we are all responsible, because we have not stood up for our brethren of the sea or fought for changes in harmful human practices. As individuals, we have made poor choices. On the other hand, there are people, organizations, and government agencies that have been paid a lot of money to steward marine resources and manage societal impacts. If the oceans were a business and its net worth had similarly decreased due to the failings of its board, CEO, or management team, heads would roll. Donald Trump would say, "Your Fired!"

Our ability to better manage human use of the sea and guide related research, education, and conservation endeavors has long been hamstrung by a lack of political and public will, insufficient funds, and conflicting priorities. And as in other high-level and high-worth issues, personal agendas, lobbying efforts, and bureaucracy are obstacles to progress. Unfortunately, as a society we tend to think on a short-term basis and address problems only when they become a crisis, as so dramatically illustrated by the recent and disastrous oil spill in the Gulf of Mexico. The oceans will always take a backseat to issues such as the economy, security, and health care. But there is an inherent link between how we manage our activities in the sea and these, the issues that determine our quality of life. It is a connection, however, that is too rarely recognized.

Humans are undeniably the most deadly predator in and influence on the oceans, and we have put the diversity of life within the sea at great risk. At human hands a mass slaughter has occurred directly through overfishing and other destructive fishing practices, and indirectly through our waste, disregard, and mismanagement of valuable natural resources. In response to our wanton ways, the oceans' resources that we have come to rely upon are now diminishing, and it is costing us jobs, food, and economic revenue. The potential for new drugs and models for biomedical research or biotechnology is also decreasing as populations decline in the sea and species become increasingly harder to find. Danger looms, and it is not just the citizens of the sea that are at risk; the well-being of humankind is also at peril.

10 The Good News

The outlook for the oceans and marine life may seem gloomy, the future doomed, an Armageddon in the making. But there is hope and even good news. As the problems that humans have created in the sea are better understood, solutions are more readily found. Across the globe people are working together to repair the damage done in the sea, to better protect its great diversity of life, and to make the use of the oceans' resources sustainable. Here are a few of the areas in which progress is being made.

CATCH OF THE DAY

The good news in fishing is that people are working together to try to make seafood a sustainable resource. Fishermen and scientists are collaborating to develop less destructive technologies, such as circle hooks, that allow more released fish to survive, and means to scare seabirds or marine mammals away from longlines or nets. Bycatch remains a serious problem, but a reduction in the loss of nontargeted species has been made through the design and use of more selective fishing gear, increased use of what is caught, and better enforcement of regulations. Management strategies are also being discovered that can allow species, such as striped bass, to rebound from overfishing. Periodic closures of fishing grounds can allow stocks, such as haddock and scallops on Georges Bank, to rebuild. Individual catch shares or quotas in some fisheries give fishermen incentives to protect their livelihoods, while preventing overfishing. In addition, to reduce the overall number of boats and people in the fishing industry, some fishermen are getting support or training so that they can shift to more sustainable revenue-generating operations, such as responsible ecotourism.

Rather than managing fisheries at the level of individual species—an approach that has proven unsound—resource managers and scientists are working to find ways to manage on an ecosystem basis. They are also trying to identify areas where spawning aggregations occur so that they can better protect them. And scientists and environmental advocates are helping local communities become better stewards of their coastal waters and the resources they provide.

Several organizations now publish best and worst choices for seafood consumption based on the best available information, some even come on a handy wallet-sized card. Three organizations that provide excellent guidance on sustainable and healthy seafood choices are the Monterey Bay Aquarium, Blue Ocean Institute, and Environmental Defense Fund (see the next chapter for their websites). The Marine Stewardship Council is an organization that now certifies sustainable fisheries, and the Seafood Choices Alliance works with chefs around the world to promote the use of sustainable seafood in cooking and restaurants. The information these groups distribute also takes into account the potential health threats from contamination by chemicals used in farming or from mercury, which comes mainly from the atmosphere and power plants and accumulates in large, predatory fishes such as tuna, sharks, swordfish, and tilefish.

In aquaculture, there is progress as well. Investment, research, and regulations are helping to improve farming operations so that they have fewer detrimental impacts, require less use of chemicals, and reduce the wild fish used in feed. One exciting area of development involves the combining of land-based fish farming with hydroponic gardens, where fish wastes are used to grow vegetables. This is especially promising for small island nations where local fisheries have been decimated and land is unsuitable for farming. Also evolving are operations in which fish farm effluents are used to grow algae, which then serve as food for cultivating shellfish or sea urchins. Researchers are working to create environmentally sound methods of offshore aquaculture as well, where habitats don't need to be destroyed and the disposal of wastes is less of a problem. Great strides have also been made in our ability to cultivate ornamental fishes for the aquarium trade. And aquariums are capturing fewer wild animals for displays and collaborating in research, rehabilitation, and captive breeding programs. These efforts provide hope that eventually we can stem the dramatic decline of fisheries due to overuse.

SAFE HAVENS

Today, very little of the ocean, less than 1 percent, is closed to fishing, yet these "no-take" areas are illustrating the great promise of this approach. Research has repeatedly shown that no-take marine reserves can lead to a rapid increase in the abundance and average size of previously exploited fishes as well as an overall rise in the diversity of marine life in the area. Marine reserves have been shown to enhance the recovery of corals in degraded reefs, and their benefits can spill over into areas beyond their borders to increase populations within adjacent waters. The idea of marine reserves is not to cordon off huge swaths of the sea so that people cannot enjoy its pleasures or fish; it is simply to protect enough area so that we can preserve biodiversity and better balance use with conservation. Creating marine reserves is part of an overall effort to use spatial planning, as we do on land, to better manage our use of and impact on the oceans. As more people recognize the benefits of marine reserves, hopefully new and larger no-take areas will be established, and not just on paper.

CLEANUP AND RESTORATION

As in other areas, there is still much to be done to clean up our wastes and restore habitats that have been destroyed. But here too there is progress. Across the globe, more people than ever before are participating in coastal cleanup days. In 2009, during an annual international cleanup organized by the Ocean Conservancy, more than four hundred thousand people volunteered across one hundred countries, collecting more than 6.8 million pounds of trash. As the impacts and scope of the marine debris issue have become more widely recognized, new programs to combat the problem have been established, and numerous groups are working to provide improved education on how to properly dispose of and recycle wastes. And more and more people are shunning the oh-so-ubiquitous plastic bags at grocery stores in favor of reusable cloth satchels.

Thousands of people have also volunteered to help restore lost habitat and marine life along our coasts—the ocean equivalent of Habitat for Humanity. Instead of building economical homes for displaced families, these people work with scientists to rebuild housing and populations in the sea. For example, in Chesapeake Bay such efforts have enabled more than 100 million native oysters to be successfully reintroduced into the region.

Along the eastern shore of Virginia, more than 200 acres of seagrass have been replanted and in Long Island's Great South Bay, 3 million native clams have been reintroduced into waters covering more than 13,000 acres. The hope in these and many other restoration projects is that once reintroduced, native species will successfully take hold and breed naturally to rebuild once plentiful populations.

CLIMATE CHANGE

Across the world people are working not only to better understand climate change, but maybe more importantly to also prepare for its impacts and, where possible, reduce the causes. Some communities are developing plans to contend with issues such as sea-level rise, increased coastal flooding and storm severity, and fisheries shifts. Some nations and individuals are investing in polices and technologies that will help to curb emissions from the burning of fossil fuels, reduce deforestation, and develop conservation strategies and alternative energies that support a more environmentally friendly way of life. The "regreening" of the planet is starting, but we have a long way to go to develop effective policies, implement them, and effectively convey the scientific evidence that overwhelmingly indicates that the Earth is warming and that humans are in large part to blame. Along these lines, scientists are working with community and global leaders to develop improved communication strategies to better inform people about climate change and its serious consequences.

INVASIVE SPECIES

Large-scale investments are now being made to combat the problem of foreign invasions in the oceans. Scientists are working with shipping operators to develop means to destroy organisms in ballast water, and regulators are trying to come up with effective and reasonable rules to reduce the impacts of ballast water release. Significant efforts are also being made to educate people about their role in helping to thwart foreign invasions by not dumping aquarium fishes in the oceans and by rinsing off small boats, trailers, fishing gear, and other equipment between trips. As for the lionfish in the Bahamas, thousands have been caught and killed, tournaments have been sponsored, rewards paid, and even new recipes developed. I hear that once you get past the poisonous spines, they are quite delicious. Local groups throughout Florida and the Caribbean have also set up quick-response teams to go out and kill lionfish once spotted. Here is one example in which "hunt to kill" is unquestionably the politically correct, environmentally friendly thing to do.

TAKING CARE OF BUSINESS

In Washington, D.C., and elsewhere there are people truly trying to make a difference. They are fighting for investment in ocean science, conservation, education, and improved management of related human activities. Some have had success, while others continue to struggle against budget shortfalls, politics as usual, and the influence of money and power. In California and Massachusetts, ocean management plans have been developed, and considerable effort is being made to put them into action, though people in both states are heavily constrained by the current state of the economy. In July 2010, President Obama signed an executive order establishing a National Policy for the Stewardship of the Ocean, Coasts, and Great Lakes. This action results from the work of several national ocean commissions and task forces going back more than six years and more than one administration. Other than making spatial planning along our coasts and in the oceans more of a priority, there is little that is actually new in this policy. What would be refreshing and absolutely encouraging, however, is if significant new funding could be acquired to implement and follow through on the recommendations made. And we could still use stronger ocean leadership in the form of an "Ocean CEO" or czar, and more champions on Capitol Hill.

EDUCATION AND ENTERTAINMENT

Today, there are more aquariums, nature parks, and opportunities for people to learn about the oceans than ever before. An estimated 135 million people visit zoos and aquariums each year, and research suggests that they come away better informed, feeling more connected to nature, and seeing themselves as part of the solution in conservation. Aquariums now host excellent education programs specifically directed at schoolchildren and provide engaging hands-on inquiry and experiential learning opportunities. In more formal settings, scientists and teachers are working together to enhance the quality and effectiveness of ocean science teaching materials and activities, and collaborative networks allow educators to share their successes.

The Internet provides new and innovative ocean education opportunities. People can go on virtual ocean expeditions, following along online as scientists explore some of the most remote and interesting areas of the sea, such as under the Arctic ice or at deep-sea hydrothermal vents. Newly discovered and bizarre creatures become instant celebrities on YouTube or blogs, and scientific articles and a wealth of information about marine life are easily accessible on the World Wide Web.

Advances in entertainment technology are enabling people to experience and see the oceans as never before. Underwater filming technology and remotely operated vehicles bring to life environments or creatures rarely seen, and high definition is allowing a sharpness and brilliance that was previously unimaginable. Television regularly allows the masses to go fishing on the high seas and to follow scientists as they study sharks and other marine life. The oceans' creatures, real or imagined, can now be seamlessly woven into popular films, even in IMAX and 3-D. Among the mesmerizing flora and fauna in James Cameron's blockbuster *Avatar* are wondrous organisms that bear a striking resemblance to those in the sea. Many of the inhabitants of the lush jungles of the imaginary planet Pandora were clearly modeled after life forms such as jellyfish, tubeworms, and sea anemones, as well as the odd headgear of the hammerhead shark, and of course the sea's dazzling nighttime displays of light—bioluminescence. The potential to experience the oceans as never before at the theater, on your television or computer, and via small mobile devices now seems almost limitless.

In the oceans, the news is not all doom and gloom; progress is being made as people across the world work to restore the sea and better protect marine life. But efforts are still too few and their scale far too small. In many places, things have changed little and the oceans are still feeling the brunt of our disregard and misuse. While the sea has proven resilient and marine life adaptable, time is running out. It will take all of us working together to change course and reverse humankind's deadly ways in the sea. The citizens of the oceans, the small, large, and even the most bizarre, are depending on us, just as we have grown to depend on them.

11 How You Can Help

As individuals and as part of a global community we can all make a difference. The power of the collective is immense, but participation even at a small scale is needed. Here are a few ways you can help the oceans and marine life.

CONSUMER CHOICES

We are a capitalistic society, driven greatly by our wants. When we demand a product in quantity, suppliers react. As a consumer, you have significant power, not only by your own choices, but also by those of your friends and colleagues. You can make smart purchasing decisions and spread the word to others.

- Choose to eat sustainable seafood, products that are derived from organisms that are not overfished, caught, or farmed using destructive methods, and that are healthy. Reliable guides to and information about sustainable seafood choices can be found at www.thefishlist.org, www.montereybay aquarium.org, www.blueoceaninstitute.org, or www.edf.org.
- If you are a chef or restaurateur, choose to use and highlight sustainable seafood products. Additional information is available at www.seafood-choices.org, www.chefscollaborative.org, and www.msc.org.
- Purchase seafood from stores and restaurants that support sustainable practices and let managers know your preference.
- Purchase organically grown food to reduce the amount of chemical fertilizers that flow into the sea.
- Don't buy jewelry or other products made of coral, seahorses, sea stars or any other organism that has been ripped from the sea.

- Purchase aquarium supplies from responsible merchants.
- Purchase eco-friendly fertilizers and use less on lawns and gardens.
- Bring recyclable bags to the market and recycle plastic bags.

CLEAN UP

Seems simple, but sometimes we forget:

- Dispose of wastes properly and recycle as much as possible.
- Encourage friends to do the same.
- Do not dump unused pharmaceuticals or chemicals down the drain.
- Wash boats after use to prevent transfer of marine organisms from one location to another.
- Don't dump aquariums or anything else into the ocean.
- Participate in a coastal cleanup. Check out local ocean advocacy groups, www.oceanconservancy.org or www.healthebay.org for information on how to get involved.

TRAVEL

Get out there and experience the sea, and do it with or through industries that support sustainability. Eco-friendly kayaking, snorkeling, diving, or whale-watching are excellent choices, as are relaxing on the beach and responsible catch-and-release fishing. When traveling, look for companies that have good environmental polices, support sustainability, and are considered eco-friendly, or ask your travel agent to do so. Two websites that might be of help are www.sustainabletravelinternational.org and www.ecotourism.org.

PARTICIPATE AND SHARE

Other ways to participate and show support for the oceans and coasts are:

- Run for an elected position that can have an influence on how public funds are spent and on how policies that impact the oceans and coasts are made.
- Choose a career or hobby that benefits the oceans and marine life.
- If you have children, encourage them to experience nature and help them to become lifelong environmental stewards.

- Visit a coastal nature center, zoo or aquarium and support their educational and conservation efforts.
- Attend local seminars and events focusing on ocean and coastal issues.
- Join a local diving, boating, or kayaking club that emphasizes eco-friendly activities.
- Slow down when boating in shallow-water areas frequented by manatees or sea turtles and take care not to damage seagrass or coral reefs.
- When possible, pick up trash along the shore or from the water.
- Encourage others to participate as well.
- Support organizations with a credible track record in ocean science, education, and conservation. If you have a specific area of interest, find out who or what organization is working on that topic and give them your support.

VOTING

Elected officials decide where we invest public funds and how we develop and use our shorelines and coastal waters. Our political representatives on planning boards, water commissions, in state legislatures, and within the federal government all play a particularly important role in how we manage our use of marine resources. We elect the school boards that oversee what educators teach and how public education funding is spent. As a voter and potential contributor to campaigns, you have tremendous power to influence who gets elected and what issues they make a priority. You can make a difference by letting your political representatives and candidates know that protecting the oceans and using our marine resources responsibly are important topics and should be more of a priority. The economy, jobs, and security weigh heavily on our minds; given the connections between these issues and the sea, focusing more on the oceans does not negate the others, but is beneficial to all.

- Phone, write, text, or email political candidates or representatives and tell them that the oceans and coasts are critically important and should be of higher priority. If you don't know the names and addresses of your congressional or senate representatives go to www.house.gov or www.senate.gov for information.
- Vote for candidates that recognize the importance of oceans and coasts to society and are working to improve our related policy and investment decisions.

- Attend political events and voice your concern over these issues.
- If a specific topic such as fisheries, waste practices, climate, or science and education is of interest to you, let your political representatives know. They truly do pay attention to what their constituents care about, especially if they hope to get re-elected.
- Let school boards know that science, the environment, and the oceans need to be better taught in our schools and that field trips to nature centers, parks, aquariums, and other places that celebrate the natural world and conservation efforts are important.
- Support organizations such as www.oceanchampions.org that try to build support for and inform our political leaders about ocean and coastal issues.
- Support policies and regulations that are good for the oceans and for marine life.

MAKING THE NEWS

In today's social networking, media-hungry, technology-driven world, information and visibility are powerful, valuable commodities. Individuals have huge influence as viewers and consumers of information, and as members of a globally connected community.

- Entertainment does not have to cater to the dimwitted or stay removed from issues of importance; it can cleverly incorporate topics of societal relevance in ways that are engaging, even funny. Let your favorite television, cable network, or film producers know that you want more substance in your programming and to include ocean-related topics. More stories with an ocean or coastal underpinning could be incorporated into popular television and film.
- Do you want to learn more and be updated about the oceans' strange creatures and their connections to your health, biotechnology, and the economy? Let the news industry know that you care about environmental and ocean issues. Many media outlets have done away with their science reporters, and few regularly do stories about the environment or ocean issues, unless a crisis looms.
- Tweet, facebook, and text your interest in the oceans and their strange marine life. Make the deadly cone snail or slime hag get more web hits than Lady Gaga. Create a funny or passionate YouTube video about the oceans and its marine life, and get it to go viral.

BIBLIOGRAPHY

Books and Journal Articles

Allen, G. R., and R. Steene. *Indo-Pacific Coral Reef Field Guide*. Singapore: Tropical Reef Research, 1994.

Allen, J. J., L. M. Mäthger, A. Barbosa, and R. T. Hanlon. "Cuttlefish use visual cues to control three-dimensional skin papillae for camouflage." *Journal of Comparative Physiology* 195 (2009): 547–555.

Barinaga, M. "Science digests the secrets of the voracious killer snails." *Science* 249, no. 4966 (1990): 250–251.

Behrens, D. W. *Nudibranch Behavior*. Jacksonville, FL: New World Publications, 2005.

Bernadsky, G., and S. E. Rosenberg. "Drag reduction of fish skin mucus: Relationship to mode of swimming and size." *Journal of Fish Biology* 42, no. 5 (1993): 797–800.

Bieri, R., and E. V. Thuesen. "The strange worm Bathybelos." *American Scientist* 78 (1990): 542–549.

Bill, R. G., and W. F. Herrnkind. "Drag reduction by formation movement in spiny lobsters." *Science* 193, no. 4258 (1976): 1146–1148.

Birkeland, C., ed. *Life and Death of Coral Reefs*. New York: Chapman & Hall, 1997.

Boles, L. C., and K. J. Lohmann. "True navigation and magnetic maps in spiny lobsters." *Nature* 421 (2003): 60–63.

Bone, Q., H. Knapp, and A. C. Pierrot-Bults, eds. *The Biology of Chaetagnaths*. Oxford: Oxford University Press, 1991.

Bonnet, D. D. "The Portugese man-of-war as a food source for the sand crab (*Emerita pacifica*)." *Science* 103, no. 2666 (1946): 148–149.

Boyle, P., and P. Rodhouse. *Cephalopods: Ecology and Fisheries*. Oxford: Blackwell Science, 2005.

Brownell, W. N., and J. M. Stevely. "The biology, fisheries, and management of the queen conch, *Stromus gigas*." *Marine Fisheries Review* 43, no. 7 (1981): 1–12.

Bruckner, A. W. "The importance of the marine ornamental reef fish trade in the wider Caribbean." *Revista de Biología Tropical* 53, suppl. 1 (2005): 127–138.

Brusca, R. C., and G. J. Brusca. *Invertebrates*. 2nd ed. Sunderland, MA: Sinauer Associates, 2002.

Bryant, D., L. Burke, J. McManus, and M. Spalding. *Reefs at Risk*. Washington, DC: World Resources Institute, USA, 1998.

Caldwell, R. L., and H. Dingle. "Stomatopods." *Scientific American* 234, no. 1 (1976): 80–89.

Caloyianis, N. "Greenland sharks." *National Geographic* 194, no. 3 (1998): 61–71.

Campbell H. A., K. P. P. Fraser, C. M. Bishop, L. S. Peck, and S. Egginton. "Hibernation in an antarctic fish: On ice for winter." *PLoS ONE* 3, no. 3 (2008): e1743. doi:10.1371/journal.pone.0001743.

Cardwell, J. R., and N. R. Liley. "Hormonal control of sex and color change in the stoplight parrotfish, *Sparisoma viride.*" *General and Comparative Endocrinology* 81, no. 1 (1991): 7–20.

Carlton, J. T. *Introduced Species in the U.S. Coastal Waters: Environmental Impacts and Management Priorities.* Arlington, VA: Pew Oceans Commission, 2001.

Caron, D. A., and N. R. Swanberg. "The ecology of planktonic sarcodines." *Aquatic Science* 3, no. 2–3 (1990): 147–180.

Chapelle, G., and L. Peck. "Polar gigantism dictated by oxygen availability." *Nature* 399 (1999): 114–115.

Chapman, D. D., M. S. Shivji, E. Louis, J. Sommer, H. Fletcher, and P. A. Prodöhl. "Virgin birth in a hammerhead shark." *Biology Letters* 3, no. 4 (2007): 425–427.

Chesher, R. H. "Destruction of Pacific corals by the sea star *Acanthaster planci.*" *Science* 165, no. 3890 (1969): 280–283.

Chivian, E., and A. Bernstein, eds. *Sustaining Life: How Human Health Depends on Biodiversity.* New York: Oxford University Press, 2008.

Chivian, E., C. M. Roberts, and A. S. Bernstein. Letter to the editor. "The threat to cone snails." *Science* 302, no. 5644 (2003): 391.

Clifton, K. E. "Mass spawning by green algae on coral reefs." *Science* 275, no. 5303 (1997): 1116–1118.

Cobb, W. D. "The Pearl of Allah." *Natural History* (November 1939): 197–202.

Cole, J., R. G. Fairbanks, and G. T. Shen. "Recent variability in the southern oscillation: isotopic results from a Tarawa Atoll coral." *Science* 260, no. 5115 (1993): 1790–1793.

Colwell, R. R. "Global climate and infectious disease: The cholera paradigm." *Science* 274, no. 5295 (1996): 2025–2031.

Colwell, R. R., and A. Huq. "Vibrios in the environment: Viable but nonculturable *Vibrio cholerae.*" In *Vibrio cholerae and Cholera: Molecular to Global Perspectives*, ed. I. K. Wachsmuth, O. Olsvik, P. A. Blake, 117–133. Washington, DC: ASM Press, 1994.

Connor, J., and C. Baxter. *Kelp Forests.* Monterey Bay, CA: Monterey Bay Aquarium Foundation, 1989.

Cooley, S. R., H. L. Kite-Powell, and S. C. Doney. "Ocean acidification's potential to alter global marine ecosystem services." *Oceanography* 22, no. 4 (2009): 172–181.

Copland, J. W., and J. S. Lucas, eds. *Giant Clams in Asia and the Pacific.* Canberra, Australia: Australian Centre for International Agricultural Research, 1988.

Corson, T. *The Secret Life of Lobsters: How Fishermen and Scientists Are Unraveling the Mysteries of Our Favorite Crustacean.* New York: HarperCollins Publishers, 2004.

Costanza, R., R. D'Arge, R. De Groot, S. Farber, M. Grasso, B. Hannon, K. Limburg, S. Naeem, R. V. O'Neil, J. Paruelo, R. G. Raskin, P. Sutton, and M. Van Den Belt. "The value of the world's ecosystem services and natural capital." *Nature* 387 (1997): 253–260.

Crist, D. T., G. Scowcroft, and J. M. Harding. *World Ocean Census*. Buffalo, NY: Firefly Books, 2009.

Cronin, T. W., and J. Marshall. "Parallel processing and image analysis in the eyes of mantis shrimps." *Biological Bulletin* 200 (2001): 177–183.

Cullon, D. L., M. B. Yunker, C. Alleyne, N. J. Dangerfield, S. O'Neill, M. J. Whiticar, and P. S. Ross. "Persistent organic pollutants in chinook salmon (*Oncorhynchus tshawytscha*): Implications for resident killer whales of British Columbia and adjacent waters." *Environmental Toxicology and Chemistry* 28, no. 1 (2009): 148–161.

DeLoach, N., and P. Humann. *Reef Creature Identification*. Jacksonville, FL: New World Publications, 2002.

———. *Reef Fish Behavior*. Jacksonville, FL: New World Publications, 2007.

———. *Reef Fish Identification*. Jacksonville, FL: New World Publications, 2008.

Demers, C., C. R. Hamdy, K. Corsi, F. Chellat, M. Tabrizian, and M. Yahia. "Natural coral exoskeleton as a bone graft substitute: A review." *Bio-medical Materials and Engineering* 12, no. 1 (2002): 15–35.

Dennett, M. R., D. A. Caron, A. F. Michaels, S. M. Gallager, and C. S Davis. "Video plankton recorder reveals high abundances of colonial Radiolaria in surface waters of the central North Pacific." *Journal of Plankton Research* 24, no.8 (2002): 797–805.

Desbruyéres, D., M. Segonzac, and M. Bright, eds. *Handbook of Deep-Sea Hydrothermal Vent Fauna*. Linz, Austria: Denisia 18, 2006.

Diaz, R. J., and R. Rosenberg. "Spreading dead zones and consequences for marine ecosystems." *Science* 321, no. 5891 (2008): 926–929.

Doney, S. C. "The growing human footprint on coastal and open-ocean biogeochemistry." *Science* 328, no. 5985 (2010): 1512–1516.

Douglas, K. "Pee is for particular." *New Scientist* 2233 (2000): 40–41.

Druehl, L. *Pacific Seaweeds: A Guide to Common Seaweeds of the West Coast*. Madeira Park, BC: Harbour Publishing, 2007.

Druffel, R. M. "Geochemistry of corals: Proxies of past ocean chemistry, ocean circulation, and climate." *Proceedings of the National Academy of Sciences* 94, no. 16 (1997): 8354–8361.

Dunbar, R. B., G. M. Wellington, M. W. Colgan, and P. W. Glynn. "Eastern Pacific sea surface temperature since 1600 A.D.: The δ18o record of climate variability in Galapagos corals." *Paleoceanography* 9, no. 2 (1994): 291–315.

Earle, S., and L. Glover. *Ocean: An Illustrated Atlas*. Washington, DC: National Geographic, 2009.

Ellis, R. *The Search for the Giant Squid*. New York: Penguin Books, 1998.

Estes, J. A., M. T. Tinker, T. M. Williams, and D. F. Doak. "Killer whale predation on sea otters linking oceanic and nearshore ecosystems." *Science* 282, no. 5388 (1998): 473–476.

Evans, P. G. H., and J. A. Raga, eds. *Marine Mammals: Biology and Conservation*. Berlin: Springer-Verlag, 2001.

Factor, J. R., ed. *Biology of the Lobster* Homarus americanus. San Diego, CA: Academic Press, 1995.

Falk, J. H., E. M. Reinhard, C. L. Vernon, K. Bronnenkant, N. L. Deans, and J. E. Heimlick. *Why Zoos & Aquariums Matter: Assessing the Impact of a Visit.* Silver Spring, MD: Association of Zoos & Aquariums, 2007.

FAO. *The State of World Fisheries and Aquaculture 2008.* Rome: FAO Fisheries and Aquaculture Department, Food and Agriculture Organization of the United Nations, 2009.

Feinberg, L. R., and W. T. Peterson. "Variability in duration and intensity of euphausiid spawning off central Oregon, 1996–2001." *Progress in Oceanography* 57, nos. 3–4 (2003): 363–379.

Fenical, W. "Marine biodiversity and the medicine cabinet: The status of new drugs from marine organisms." *Oceanography* 9, no. 1 (1996): 23–27.

Fish, F. E. "Wing design and scaling of flying fish with regard to flight performance." *Journal of Zoology (London)* 221 (1990): 391–403.

Ford, J. K. B., and G. M. Ellis. "Selective foraging by fish-eating killer whales *Orcinus orca* in British Columbia." *Marine Ecology Progress Series* 316 (2006): 185–199.

Fry, B. G., K. Roelants, and J. A. Norman. "Tentacles of Venom: Toxic protein convergence in the kingdom Animalia." *Journal of Molecular Evolution* 68, no. 4 (2009): 311.

Garm, A., and S. Mori. "Multiple photoreceptor systems control the swim pacemaker activity in box jellyfish." *Journal of Experimental Biology* 212 (2009): 3951–3960.

Gilmer, R. W. "In situ observations of feeding behavior of thecosome pteropod molluscs." *American Malacological Bulletin* 8, no. 1 (1990): 53–59.

Gilmore, R. G., J. W. Dodrill, and P. A. Linley. "Reproduction and embryonic development of the sand tiger shark, *Odontaspis taurus (Rafinsque)." *Fishery Bulletin* 81, no. 2 (1983): 201–225.

Gowell, E. *Amazing Jellies: Jewels of the Sea.* Piermont, NH: Bunker Hill Publishing Co. and New England Aquarium, 2004.

Graham, M. H., B. P. Kinlan, L. D. Druehl, L. E. Garsek, and S. Banks. "Deep-water kelp refugia as potential hotspots of tropical marine diversity and productivity." *Proceedings of the National Academy of Sciences* 104, no. 42 (2007): 16576–16580.

Guest, J. R., A. H. Baird, K. E. Clifton, and A. J. Heyward. "From molecules to moonbeams: Spawning synchrony in coral reef organisms." *Invertebrate Reproduction and Development* 51, no. 3 (2008) 145–149.

Haag, A. "Whale fall." *Nature* 433 (2005): 566–567.

Hall, K. C., and R. T. Hanlon. "Principal features of the mating system of a large spawning aggregation of the giant Australian cuttlefish *Sepia apama* (Mollusca: Cephalopoda)." *Marine Biology* 140 (2002): 533–545.

Hamner, W. M. "Australia's box jellyfish: A killer down under." *National Geographic* 186, no. 2 (1994): 116–130.

Hanlon, R. T. "Cephalopod dynamic camouflage." *Current Biology* 17, no. 11 (2007): R400–R404.

Hanlon, R. T., and J. B. Messenger. *Cephalopod Behavior*. Cambridge: Cambridge University Press, 1996.

Hanlon, R. T., M. J. Naud, P. W. Shaw, and J. N. Havenhand. "Behavioural ecology: Transient sexual mimicry leads to fertilization." *Nature* 430 (2005): 212.

Haq, B. U., and A. Boersma. *Introduction to Marine Micropaleontology*. New York: Elsevier Science Publishing, 1983.

Hargens, A. R., and S. V. Shabica. "Protection against lethal freezing temperatures by mucus in an Antarctic limpet." *Cryobiology* 10, no. 4 (1973): 331–337.

Hay, M. "Enhanced: Synchronous spawning—When timing is everything." *Science* 275, no. 5303 (1997): 1080–1081.

Hay, M. E., V. J. Paul, S. M. Lewis, K. Gustafson, J. Tucker, and R. N. Trindell. "Can tropical seaweeds reduce herbivory by growing at night? Diel patterns of growth, nitrogen content, herbivory, and chemical versus morphological defenses." *Oecologia* 75, no. 2 (2004): 233–245.

Heithaus, M. R., A. Frid, A. J. Wirsing, and B. Worm. "Predicting ecological consequences of marine top predator declines." *Trends in Ecology and Evolution* 23, no. 4 (2008): 202–210.

Helfman, G. S., B. B. Collette, D. E. Facey, and B. W. Bowen. *The Diversity of Fishes*. West Sussex, U.K.: Wiley-Blackwell, 2009.

Helvarg, D. *50 Ways to Save the Ocean*. Makawao, HI: Inner Ocean Publishing, 2006.

Herren, L. W., L. J. Walters, and K. S. Beach. "Fragment generation, survival and attachment of Dictyota species at Conch Reef in the Florida Keys, USA." *Coral Reefs* 25 (2006): 287–295.

Herrnkind, W. "Queuing behavior of spiny lobsters." *Science* 164, no. 3886 (1969): 1425–1427.

Hoegh-Guldberg, O., and J. F. Bruno. "The impact of climate change on the world's marine ecosystems." *Science* 328, no. 5985 (2010): 1523–1528.

Hoelzel, A. R. "Killer whale predation on marine mammals at Punta Norte, Argentina: Food sharing, provisioning and foraging strategy." *Proceedings of the Royal Society Bulletin* 269 (1991): 1467–1475.

Hogg, C. J., T. L. Rogers, A. Shorter, K. Barton, P. J. O. Miller, and D. Nowacek. "Determination of steroid hormones in whale blow: It is possible." *Marine Mammal Science* 25, no. 3 (2009): 605–618.

Hudson, J. H., E. A. Shinn, R. B. Halley, and B. Lidz. "Sclerochronology: A tool for interpreting past environments." *Geology* 4, no. 6 (1976): 361–364.

International Seakeepers Society. *11 Critical Ocean Issues: With Action Items for Concerned Citizens*. Ft. Lauderdale, FL: International Seakeepers Society, 2006.

Isdale, P. "Fluorescent bands in massive corals record centuries of coastal rainfall." *Nature* 310 (1984): 578–579.

Jackson, J. B. C., M. X. Kirby, W. H. Berger, K. A. Bjorndal, L. W. Botsford, B. J. Bourque, R. H. Bradbury, R. Cooke, J. Erlandson, J. A. Estes, T. P. Hughes, S. Kidwell, C. B. Lange, H. S. Lenihan, J. M. Pandolfi, C. H. Peterson, R. S. Steneck, M. J. Tegner, and R. R. Warner. "Historical overfishing and the recent collapse of coastal ecosystems." *Science* 293 (2001): 629–638.

Jeffryes, C., T. Gutu, J. Jiao, and G. L. Rorrer. "Metabolic insertion of nanostructured TiO_2 into the patterned biosilica of the diatom *Pinnularia* sp. by a two-stage bioreactor cultivation process." *ACS Nano* 2, no. 10 (2008): 2103–2112.

Jensen, D. "The hagfish." *Scientific American* 214, no. 2 (1966): 82–90.

Johnson, S., E. J. Balsert, and E. A. Widder. "Light emitting suckers in an octopus." *Nature* 398 (1999): 113.

Jones, E. C. "*Tremoctopus violaceus* uses *Physallia* tentacles as weapons." *Science* 139, no. 3556 (1963): 764–766.

Jorgensen, J. M., J. P. Lomholt, R. E. Weber, and H. Malte, eds. *The Biology of Hagfishes*. London: Chapman & Hall, 1998.

Kaiser, J. "The Dirt on ocean garbage patches." *Science* 328, no. 5985 (2010): 1506.

Kanciruk, P., and W. F. Herrnkind. "Mass migration of spiny lobster *Panulirus argus* (Crustacea: Palinuridae): Behaviour and environmental correlates." *Bulletin of Marine Science* 28 (1978): 601–623.

Kelleher, K. "Discards in the world's marine fisheries: An update." *FAO Fisheries Technical Paper*, no. 470 (2005).

Knop, D. *Giant clams: A Comprehensive Guide to the Identification and Care of Tridacnid Clams*. Coconut Grove, FL: Ricordea Publishing, 1996.

Knox, G. A. *The Biology of the Southern Ocean*. Cambridge: Cambridge University Press, 1994.

Koehl, M. A. R., and J. R. Strickler. "Copepod feeding currents: Food capture at low Reynolds number." *Limnology and Oceanography* 26, no. 6 (1981): 1062–1073.

Kubodera, T., and K. Mori "First-ever observations of a live giant squid in the wild." *Proceedings of the Royal Society B* 272 (2005): 2583–2586.

Lalli, C. M., and R. W. Gilmer. *Pelagic Snails: The Biology of Holoplanktonic Gastropod Mollusks*. Stanford: Stanford University Press, 1989.

Lambert, P. *Sea Stars of British Columbia, SE Alaska, and Puget Sound*. Vancouver: UBC Press, 2000.

Land, M. F. "The spatial resolution of the pinhole eyes of giant clams (Tridacna maxima)." *Proceedings of the Royal Society (London)* 270 (2003): 185–188.

Lane, C. E. "The Portugese man-of-war." *Scientific American* 202, no. 3 (1960): 158–168.

Lang, J. "Interspecific aggression by scleractinian corals." *Bulletin of Marine Science* 23 (1973): 260–279.

Lewis, C., and T. A. F. Long. "Courtship and reproduction in *Carybdea sivickisi*." *Marine Biology* 147 (2005): 477–483.

Lucas, J. S. "The biology, exploitation and mariculture of giant clams (*Tridacnidae*)." *Reviews in Fisheries Science* 2, no. 3 (1994): 181–223.

Madin, L. P., and C. M. Cetta. "The use of gut fluorescence to estimate grazing by oceanic salps." *Journal of Plankton Research* 6, no. 3 (1984): 475–492.

Madin, L. P., P. Kremer, P. H. Wiebe, J. E. Purcell, E. H. Horgan, and D. A. Nemazie. "Periodic swarms of the salp *Salpa aspera* in the slope water off the NE United States: Biovolume, vertical migration, grazing, and vertical flux." *Deep Sea Research Part I: Oceanographic Research Papers* 53, no. 5 (2006): 804–819.

Mann, D. A., J. V. Locasciol, F. C. Coleman, and C. C. Koenig. "Goliath grouper *Epinephelus itajara* sound production and movement patterns on aggregation sites." Endangered Species Research, published online October 10, 2008, at http://www.int-res.com/articles/esr2009/7/n007p229.pdf.

Mann, J., R. C. Connor, P. L. Tyack, and H. Whitehead, eds. *Cetacean Societies: Field Studies of Dolphins and Whales*. Chicago: University of Chicago Press, 2000.

Marris, E. "Marine natural products: Drugs from the deep." *Nature* 443 (2006): 904–905.

Martini, F. H. "Secrets of the slime hag." *Scientific American* (October 1998): 70–75.

Mate, B. R., P. Duley, B. A. Lagerquist, F. Wenzel, A. Stimpert, and P. Clapham. "Observations of a female North Atlantic right whale (*Eubalaena glacialis*) in simultaneous copulation with two males: Supporting evidence for sperm competition." *Aquatic Mammals* 31 (2005): 157–160.

Mäthger, L. M., A. Barbosa, S. Miner, and R. Hanlon. "Color blindness and contrast perception in cuttlefish (*Sepia officinalis*) determined by a visual sensorimotor assay." *Vision Research* 46 (2006): 1746–1753.

Mauchline, J., and L. R. Fisher. "The biology of euphausiids." *Advances in Marine Biology* 7, New York: Academic Press, 1969.

Mauzey, K. P., C. Birkeland, and P. K. Dayton. "Feeding behavior of asteroids and escape response of their prey in the Puget Sound region." *Ecology* 49, no. 4 (1968): 603–619.

McClintock, J. B., R. A. Angus, C. P. Ho, C. D. Amsler, and B. J. Baker. "Intraspecific agonistic arm-fencing behavior in the Antarctic keystone sea star *Odontaster validus* influences prey acquisition." *Marine Ecology Progress Series* 371 (2008): 297–300.

McClintock, J. B., and J. Janssen. "Pteropod abduction as a chemical defense in a pelagic Antarctic amphipod." *Nature* 346 (1990): 462–464.

McClintock, J. B., D. P. Swenson, D. K. Steinberg, and A. A. Michaels. "Feeding deterrent properties of common holoplankton from Bermuda waters." *Limnology and Oceanography* 41, no. 4 (1996): 798–801.

McComb, D. M., T. C. Tricas, and S. M. Kajiura. "Enhanced visual fields in hammerhead sharks." *Journal of Experimental Biology* 212 (2009): 4010–4018.

McWhinnie, M. A., and C. J. Denys. "The high importance of the lowly krill." *Natural History* 89, no. 3 (1980): 66–73.

Meinesz, A. *Killer Algae: The True Tale of a Biological Invasion*. Chicago: University of Chicago Press, 1999.

Moniz, R. "Interview: Dr. Roger Hanlon." *Fathoms* 20 (2007): 92–105.

Monterey Bay Aquarium. "The mystifying clear headed fish." *Ocean Geographic* 8 (2009): 11–12.

Morrissey, J. F., and J. L. Sumich. *Introduction to the Biology of Marine Life*. 9th ed. Sudbury, MA: Jones and Bartlett Publishers, 2009.

Mumby, P. J., and A. R. Harborne. "Marine reserves enhance the recovery of corals on Caribbean reefs." *Public Library of Science* 5, no. 1 (2010): e8657. doi:10.1371/journal.pone.0008657.

Myers, R. A., J. K. Baum, T. D. Shepherd, S. P. Powers, and C. H. Peterson. "Cascading effects of the loss of apex predatory sharks from a coastal ocean." *Science* 315, no. 5820 (2007): 1846–1850.

Myers, R. A., and B. Worm. "Rapid worldwide depletion of predatory fish communities." *Nature* 423 (2003): 280–283.

National Research Council. *Oil in the Sea: Inputs, Fates and Effects.* Washington, DC: National Academy Press, 2003.

Naud, M., R. T. Hanlon, K. Hall, P. W. Shaw, and J. N. Havenhand. "Behavioural and genetic assessment of reproductive success in a spawning aggregation of the Australian giant cuttlefish, *Sepia apama.*" *Animal Behavior* 67, no. 6 (2004): 1043–1050.

Nilsson, D. E., L. Gislen, M. M. Coates, C. Skogh, and A. Garm. "Advanced optics in a jellyfish eye." *Nature* 435 (2005): 201–204.

Noad, M. J., D. H. Cato, M. M. Bryden, M. N. Jenner, and K. C. S. Jenner. "Cultural revolution in whale song." *Nature* 408 (2000): 537.

Norman, M. D., D. Paul, J. Finn, and T. Tregenza. "First encounter with a live male blanket octopus: The world's most sexually size-dimorphic large animal." *New Zealand Journal Marine and Freshwater Research* 36 (2002): 733–736.

Norris, K. S. "White whale of the North: Beluga." *National Geographic* 185, no. 6 (1994): 2–31.

Nouvian, C. *The Deep: The Extraordinary Creatures of the Abyss.* Chicago: University of Chicago Press, 2007.

O'Dor, R. K., C. O. Parkes, and D. H. Copp. "Amino acid composition of salmon calcitonin." *Canadian Journal of Biochemisty* 47, no. 8 (1969): 823–825.

Olivera, B. M., J. Rivier, C. Clark, C. A. Ramilo, G. P. Corpuz, F. C. Abogadie, E. E. Mena, S. R. Woodward, D. R. Hillyard, and L. J. Cruz. "Diversity of *Conus* neuropeptids." *Science* 249, no. 4966 (1990): 257–263.

Orrl, J. C., V. J. Fabry, O. Aumont, L. Bopp, S. C. Doney, R. A. Feely, A. Gnanadesikan, N. Gruber, A. Ishida, F. Joos, R. M. Key, K. Lindsay, E. Maier-Reimer, R. Matear, P. Monfray, A. Mouchet, R. G. Najjar, G. Plattner, K. B. Rodgers, C. L. Sabine, J. L. Sarmiento, R. Schlitzer, R. D. Slater, I. J. Totterdell, M. Weirig, Y. Yamanaka, and A. Yool. "Anthropogenic ocean acidification over the twenty-first century and its impact on calcifying organisms." *Nature* 437 (2005): 681–686.

Parker, S. *The Encyclopedia of Sharks.* Buffalo: Firefly Books, 2008.

Pepperell, J. *Fishes of the Open Ocean: A Natural History & Illustrated Guide.* Davie, FL: Guy Harvey, 2010.

———. "Faster, higher." *Bluewater Boats and Sportfishing* (October/November 2005): 58–62.

Perrin, W. F., B. Wursig, and J. G. M. Thewissen, eds. *Encyclopedia of Marine Mammals.* 2nd ed. London: Academic Press, 2008.

Perrine, D. "Mysterious predators of the frozen North." *Ocean Geographic* 8 (2009): 39–49.

Pietsh, T. W., and D. B. Grobecker. "The complete angler: Aggressive mimicry in an antennariid anglerfish." *Science* 201, no. 4353 (1978): 369–370.

Pincebourde, S., E. Sanford, and B. Helmuth. "An intertidal sea star adjusts thermal inertia to avoid extreme body temperatures." *American Naturalist* 174, no. 6 (2009): 890–897.

Prager, E. *Chasing Science at Sea: Racing Hurricanes, Stalking Sharks, and Living Undersea with Ocean Experts.* Chicago: University of Chicago Press, 2008.

———. *The Oceans.* New York: McGraw-Hill, 2000.

Ralls, K. "Sperm competition in grey whales." *Nature* 336 (1998): 116–117.

Randall, J. E. "Contributions to the biology of the queen conch, *Strombas gigas.*" *Bulletin of Marine Science of the Gulf and Caribbean* 14, no. 2 (1964): 246–295.

———. "The habits of the queen conch." *Sea Frontiers* 10, no. 3 (1964): 230–239.

———. "Monarch of the grass flats." *Sea Frontiers* 9, no. 3 (1963): 160–167.

Richardson, C. A., P. Dustan, and J. C. Lang. "Maintenance of living space by sweeper tentacles of *Montastrea cavernosa*, a Caribbean reef coral." *Marine Biology* 55 (1979): 181–186.

Roberts, C. M., J. A. Bohnsak, F. Gell, J. P. Hawkings, and R. Goodridge. "Effects of marine reserves on adjacent fisheries." *Science* 294 (2001): 1920–1923.

Robison, B. H. "Light in the ocean midwater." *Scientific American* 273 (1995): 60–64.

Robison, B. H., K. R. Reisenbichler, J. C. Hunt, and S. H. D. Haddock. "Light production by the arm tips of the deep-sea cephalopod *Vampyroteuthis infernalis.*" *Biological Bulletin* 205 (2003): 102–109.

Rommel, S. A., A. M. Costidis, A. Fernandez, P. D. Jepson, D. A. Pabst, W. A. Mclellan, D. S. Houser, T. W. Cranford, A. L. Vanhelden, D. M. Allen, and B. Barros. "Elements of beaked whale anatomy and diving physiology and some hypothetical causes of sonar-related stranding." *Journal of Crustacean Research and Management* 7, no. 3 (2006): 189–209.

Rosen, M. W., and N. E. Cornford. "Fluid friction of fish slimes." *Nature* 234 (1971): 49–51.

Rouse, G. W., S. K. Goffredi, and R. C. Vrijenhoek. "Osedax: Bone-eating marine worms with dwarf males." *Science* 305, no. 5684 (2004): 668–671.

Roux, F. X., D. Brasnu, B. Loty, B. George, and G. Fuillemin. "Madreporic coral: A new bone graft substitute for cranial surgery." *Journal of Neurosurgery* 69 (1988): 510–513.

Ruggiero, G. "The giant clam: Friend or foe?" *Sea Frontiers* 31, no. 1 (1985): 4–9.

Rumpho, M. E., J. M Worful, J. Lee, K. Kannan, M. S. Tyler, D. Bhattacharya, A. Moustafa, and J. R. Manhart. "Horizontal gene transfer of the algal nuclear gene *psbO* to the photosynthetic sea slug *Elysia chlorotica.*" *Proceedings of the National Academy of Science* 105, no. 46 (2008): 17867–17871.

Santelices, B. "The discovery of kelp forests in deep-water habitats of tropical regions." *Proceedings of the National Academy of Science* 104, no. 49 (2007): 19163–19164.

Scales, H. *Poseidon's Steed.* New York: Penguin Group (USA), 2009.

Schofield, O., H. W. Ducklow, D. G. Martinson, M. P. Meredith, M. A. Moline, and W. R. Fraser. "How do polar marine ecosystems respond to rapid climate change?" *Science* 328, no. 5985 (2010): 1520–1523.

Schusterman, R. J., D. Kastak, D. H. Levnson, C. J. Reichmuth, and B. L. Southall. "Why pinnipeds don't echolocate." *Journal of the Acoustical Society of America* 107, no. 4 (2000): 2256–2264.

Selig, E. R., and J. F. Bruno. "A global analysis of the effectiveness of marine protected areas in preventing coral loss." *Public Library of Science* 5, no. 2 (2010): e9278. doi:10.1371/journal.pone.0009278.

Shinn, E. A. "Time capsules in the sea." *Sea Frontiers* 27, no. 6 (1981): 364–374.

Siegel, D. A., and B. A. Franz. "Oceanography: Century of phytoplankton change." *Nature* 466, no. 7306 (2010): 569–571.

Smith, D. L., and K. B. Johnson. *A Guide to Marine Coastal Plankton and Marine Invertebrate Larvae*. 2nd ed. Dubuque, IA: Kendall/Hunt Publishing, 1996.

Sotka, E. E., M. E. Hay, and J. D. Thomas. "Host-plant specialization by a non-herbivorous amphipod: Advantages for the amphipod and costs for the seaweed." *Oecologia* 118, no. 4 (1999): 471–482.

Soto, N. A., M. P. Johnson, P. T. Madsen, F. Diaz, I. Dominguez, A. Brito, and P. Tyack. "Cheetahs of the deep sea: Deep foraging sprints in short-finned pilot whales off Tenerife (Canary Islands)." *Journal of Animal Ecology* 77, no. 5 (2008 online): 936–947.

Southwell, M. W., J. B. Weisz, C. S. Martens, and N. Lindquist. "In situ fluxes of dissolved inorganic nitrogen from the sponge community on Conch Reef, Key Largo, Florida." *Limnology and Oceanography* 53, no. 3 (2008): 986–996.

Southwick Associates. *Sportfishing in America: An Economic Engine and Conservation Powerhouse*. N.p.: Produced for the American Sportfishing Association with funding from the Multistate Conservation Grant Program, 2007.

Steger, R., and R. L. Caldwell. "Intraspecific deception by bluffing: A defense strategy of newly molted stomatopods (Arthropoda: Crustacea)." *Science* 221, no. 4610 (1983): 558–560.

Stoner, A. W., and M. Ray-Culp. "Evidence for the Allee effects in an over-harvested marine gastropod: Density-dependent mating and egg production." *Marine Ecology Progress Series* 202 (2000): 297–302.

Taylor, M. W., R. Radax, D. Steger, and M. Wagner. "Sponge-associated microorganisms: Evolution, ecology, and biotechnological potential." *Microbiology and Molecular Biology Reviews* 71, no. 2 (2007): 295–347.

Thomas, R. F. "Systematics, distribution, and biology of cephalopods of the genus *Tremoctopus*." *Bulletin of Marine Science* 27, no. 3 (1977): 353–392.

Thomson, R. E., S. F. Mihály, A. B. Rabinovich, R. E. McDuff, S. R. Veirs, and F. R. Stahr. "Constrained circulation at Endeavour ridge facilitates colonization by vent larvae." *Nature* 424 (2003): 545–549.

Thuesen, E. V., and K. Kogure. "Bacterial production of tetrodotoxin in four species of chaetagnatha." *Biological Bulletin* 176 (1989): 191–194.

Tricas, T. C., K. Deacon, P. Last, J. E. McCosker, T. I. Walker, and L. Taylor. *Sharks & Rays*. Sydney, Australia: Time-Life Books, 1997.

Tyack, P. L., M. Johnson, N. A. Soto, A. Sturlese, and P. T. Madsen. "Extreme diving behavior of beaked whale species *Ziphius cavirostris* and *Mesoplodon densirostris*." *Journal of Experimental Biology* 209 (2006): 4238–4253.

U.S. Commission on Ocean Policy. *An Ocean Blueprint for the 21st Century.* Final Report. Washington, DC: U.S. Commission on Ocean Policy, 2004.

Van Dover, C. *The Ecology of Deep-Sea Hydrothermal Vents.* Princeton, NJ: Princeton University Press, 2000.

Van Dover, C., C. R. German, K. G. Speer, L. M. Parson, And R. C. Vrijenhoek. "Evolution and biogeography of deep-sea vent and seep invertebrates." *Science* 295, no. 5558 (2002): 391–403.

Videler, J. J. *Fish Swimming.* London: Chapman and Hall, 1993.

Vogel, G. "Europe tries to save its eels." *Science* 329, no. 5991 (2010): 505–507.

Vogel, G. "The inner lives of sponges." *Science* 320, no. 5879 (2008): 1028–1030.

Waller, G., ed. *SeaLife: A Complete Guide to the Marine Environment.* With M. Burchett and M. Dando. Washington, DC: Smithsonian Institution Press, 1996.

Wallis, C. *Seahorses: Mysteries of the Oceans.* Boston: Bunker Hill Publishing, 2004.

Walls, J. G. *Cone Shells: A Synopsis of the Living Conidae.* Neptune City, NJ: T. F. H. Publications, 1979.

Walters, L., and K. Beach. "Algal bloom in the Florida Keys." *Underwater Naturalist* 25, no. 3 (2001): 27–29.

Wild, C., M. Huettel, A. Klueter, S. G. Kremb, M. Y. M. Rasheed, and B. B. Jorgensen. "Coral mucus functions as an energy carrier and particle trap in the reef ecosystem." *Nature* 428 (2004): 66–70.

Wilkinson, C., ed. "Status of coral reefs of the world: 2008." Townsville, Australia: Global Coral Reef Monitoring Network and Reef and Rainforest Research Center, 2008.

Williams, T. M., R. W. Davis, L. A. Fuiman, J. Francis, B. J. Le Boeuf, M. Horning, J. Calambokidis, and D. A. Croll. "Sink or swim: Strategies for cost-efficient diving by marine mammals." *Science* 288 no. 5463 (2000): 133–136.

Weis, V. M., and D. Allemand. "What determines coral health?" *Science* 324, no. 5931 (2009): 1153–1155.

Wood, E. M. *Collection of Coral Reef Fish for Aquaria: Global Trade, Conservation Issues and Management Strategies.* Ross-on-Wye, Herefordshire, UK: Marine Conservation Society, 2001.

Woodhead, A. D. *Nonmammalian Animal Models for Biomedical Research.* Boca Raton, FL: CRC Press, 1990.

Worm, B., M. Sanlow, A. Oschlies, H. K. Lotze, and R. A. Myers. "Global patterns of predator diversity in the open oceans." *Science* 309, no. 5739 (2005): 1365–1369.

Yonge, C. M. "Giant clams." *Scientific American* 232, no. 4 (1975): 96–105.

Online References

3Dchem.com. "Chitosan." http://www.3dchem.com/molecules.asp?ID=444.

ABC.net.au. "Coral sunscreen finally sees the light." *News in Science.* http://www.abc.net.au/science/news/stories/s102327.htm.

Alaska Whale Foundation. "Bubble net feeding." http://www.alaskawhalefoundation.org/education/bubble_net/bubble_net_feeding.html.

American Cetacean Society. http://www.acsonline.org.

American Cetacean Society. Puget Sound Chapter. http://www.acspugetsound
.org/facts/index.html.

American Sportfishing Association. http://www.asafishing.org/statistics/index.html.

Antarctic and Southern Ocean Coalition. http://www.asoc.org/.

Australian Antarctic Division. http://www.aad.gov.au.

Australian Museum. "Sea slug forum." http://www.seaslugforum.net.

BBC News. "Fast flying fish glides by ferry." http://news.bbc.co.uk/2/hi/7410421
.stm?lsm.

BBC News. "Starfish pump up to cool down." http://news.bbc.co.uk/earth/hi/earth_
news/newsid_8328000/8328311.stm.

BBC News. "Strange jellies of the icy depths." http://news.bbc.co.uk/earth/hi/earth_
news/newsid_8231000/8231367.stm.

BBC News. "Zombie worms found off Sweden." http://news.bbc.co.uk/2/hi/science/
nature/4354286.stm.

The Billfish Foundation. http://www.billfish.org.

California Academy of Science. "Extreme life." http://www.calacademy.org/exhibits/
xtremelife/life_on_earth.php.

California Department of Fish and Game. "Giant kelp." http://www.dfg.ca.gov/
marine/status/report2003/giantkelp.pdf.

Carl Von Ossietzky University Oldenburg. "Biology of copepods." Department of
Zoosystematics and Morphology. http://www.uni-oldenburg.de/zoomorphology/
Biology.html.

Census of Marine Life. http://www.coml.org.

The Cephalopod Page. "20,000 tentacles under the sea—Cephalopods in cinema."
http://www.thecephalopodpage.org/.

Chemistry World. "A glowing green Nobel." http://www.rsc.org/images/Nobel_
tcm18-137088.pdf.

Chinese-Food Recipes.com. "Sea cucumber in Chinese cooking." http://www
.chinesefood-recipes.com/food_articles/sea_cucumber_chinese_cooking.php.

ConeShell.net. http://www.coneshell.net.

Coralscience.org. http://www.coralscience.org.

CoralScience.org. "The science behind stomatopods." http://www.coralscience
.org/main/articles/reef-species-4/stomatopods.

Council on Environmental Quality. "The interacgency policy task force." http://www
.whitehouse.gov/administration/eop/ceq/initiatives/oceans.

Darguard. "Antarctic penguins." http://www.gdargaud.net/Antarctica/Penguins.html.

Dive Training. "The quick and the deadly." http://www.dtmag.com/Stories/
Marine%20Life/10-05-whats_that.htm.

The Echinoblog. http://echinoblog.blogspot.com.

Environmental Health News. "Salmon in near-shore Pacific contaminating killer
whales." http://www.environmentalhealthnews.org/ehs/news/contaminated-
killer-whales.

Fishupdate.com. "Scotland: Starfish slime is liquid asset for new biotech company."

http://www.fishupdate.com/news/fullstory.php/aid/3303/Scotland:_Starfish_
slime_is_liquid_asset_for_new_biotech_company.html.

Food and Agriculture Organization of the United Nations. "Fisheries." http://www
.fao.org/fishery/en.

Food and Agriculture Organization of the United Nations. "Marine lobsters of the
world." http://nlbif.eti.uva.nl/bis/lobsters.php?menuentry=inleiding.

Federal Trade Commission. "FTC and FDA take new action in fight against deceptive
marketing." http://www.ftc.gov/opa/2003/06/trudeau.shtm.

Fortune Magazine. "Fortune 500: Our annual ranking of America's largest corpora-
tions." http://money.cnn.com/magazines/fortune/fortune500/2010/full_list/.

Frogfish. http://www.frogfish.ch.

Gaia Discovery. "Medicines from coral reefs." http://www.gaiadiscovery.com/marine-
life-latest/medicines-from-coral-reefs.html.

The Greenland Shark and Elasmobranch Education and Research Group. http://www
.geerg.ca.

Harbor Branch Oceanographic Institution. "Marine biotech." http://www
.marinebiotech.org/.

Harbor Branch Oceanographic Institution, Aquaculture Division. "Species profile:
Queen conch, *Strombus gigas.*"http://www.ca.uky.edu/wkrec/QueenConch.pdf.

International Association for Bear Research and Management. "Polar bear." http://
www.bearbiology.com/iba/bears-of-the-world/polar-bear.html.

International Fund for Animal Welfare. "Whale watching worldwide: Tourism
numbers, expenditures, and expanding economic benefits." http://www.ifaw
.org/whalewatchingworldwide.

The JelliesZone. http://jellieszone.com/.

Journey North. "Gray whales." http://www.learner.org/jnorth/tm/gwhale/About-
Spring.html.

LiveScience.com. "Helicopters collect whale snot from blowholes." http://www
.livescience.com/animals/081113-whale-spit.html.

The Lobster Conservancy. http://www.lobsters.org.

Marine Biological Laboratory. "Laboratory of Roger Hanlon." http://www.mbl.edu/
mrc/hanlon/index.html.

Marinebio.net. "Migration of the gray whale." http://www.marinebio.net/
marinescience/05nekton/GWmigration.htm.

Marinebio.net. "The walrus." http://www.marinebio.net/marinescience/04benthon/
arcwalrus.htm.

Mass High Tech: The Journal of New England Technology. "Engineers build robotuna
for Navy." http://www.masshightech.com/stories/2008/07/21/weekly15-Engineers-
build-RoboTuna-for-Navy.html.

Monterey Bay Aquarium. "Seafood watch." http://www.montereybayaquarium
.org/cr/cr_seafoodwatch/sfw_recommendations.aspx.

Monterey Bay Research Institute. "Researchers solve mystery of deep-sea fish with
tubular eyes and transparent head." http://www.mbari.org/news/news_
releases/2009/barreleye/barreleye.html.

Monterey Bay Research Institute. "Understanding benthic biodiversity: The curious case of bone-eating worms." http://www.mbari.org/news/publications/ar/chapters/08_BenthicBiodiversity.pdf.

MSNBC.com. "Sea-sponge breast cancer drug may extend lives." http://www.msnbc.msn.com/id/37540359/ns/health-cancer/.

MSNBC.com. "The bizarre case of bone-eating worms." http://www.msnbc.msn.com/id/5549064/ns/technology_and_science-science/.

National Geographic News. "'Walnut-size' male octopus seen alive for first time." http://news.nationalgeographic.com/news/2003/02/0212_030212_walnutoctopus.html.

National Oceanic and Atmospheric Administration. "Marine Debris Program." http://www.marinedebris.noaa.gov.

National Oceanic and Atmospheric Administration. "Ocean explorer." http://oceanexplorer.noaa.gov/.

National Oceanic and Atmospheric Administration, National Marine Fisheries Service. "By the numbers: Saltwater fishing facts and figures for 2006." http://www.nmfs.noaa.gov/sfa/PartnershipsCommunications/recfish/BytheNumbers2006.pdf.

National Oceanic and Atmospheric Administration, National Marine Fisheries Service. "Fishwatch: U.S. seafood facts." http://www.nmfs.noaa.gov/fishwatch/.

National Oceanic and Atmospheric Administration, National Marine Fisheries Service. "Sustainability: It's in our hands." http://www.nmfs.noaa.gov/speciesid/Sustainability.html.

National Oceanic and Atmospheric Administration, National Ocean Service. "Corals." http://oceanservice.noaa.gov/education/kits/corals/welcome.html.

National Resource Center for Cephalopods. http://www.cephalopod.org.

National Science Foundation. "Jellyfish gone wild." http://www.nsf.gov/news/special_reports/jellyfish/biology.jsp.

New York Times. "Watching whales watching us." http://www.nytimes.com/2009/07/12/magazine/12whales-t.html.

New York Times. "Doctors trying coral for skeletal repairs." http://www.nytimes.com/1991/07/02/health/doctors-trying-coral-for-skeletal-repairs.html.

The Ocean Conservancy. http://www.oceanconservancy.org.

Oceansunfish.org. "The ocean sunfish." http://www.oceansunfish.org/.

Pearl-Guide.com. "The pearl of Allah: The facts, the fiction, and the fraud." http://www.pearl-guide.com/the-pearl-of-allah.shtml.

Penguin Science. http://www.penguinscience.com/index.php.

Pennsylvanian State University. "New worms discovered by professor." http://www.collegian.psu.edu/archive/1997/07/07-31-97tdc/07-31-97d01-002.htm.

Polar Bear International. http://www.polarbearsinternational.org.

Project Seahorse. http://seahorse.fisheries.ubc.ca/.

Queensland Museum. "Cone shells." http://www.qm.qld.gov.au/inquiry/factsheets/cone_shells_20080709.pdf.

Radiolaria.org. http://www.radiolaria.org/.

ReefQuest Centre for Shark Research. "Biology of sharks and rays." http://www
.elasmo-research.org/.

ScienceCentral.org. "War bandages." http://www.sciencentral.com/articles/view
.php3?type=article&article_id=218392341.

ScienceDaily. "All octopuses are venomous: Could lead to drug discovery." http://
www.sciencedaily.com/releases/2009/04/090415102215.htm.

ScienceDaily. "Seaweed to tackle the rising tide of obesity." http://www.sciencedaily
.com/releases/2010/03/100321203508.htm.

ScienceDaily. "'Unicorn' whale's 8-foot tooth discovered." http://www.sciencedaily
.com/releases/2005/12/051214081832.htm.

Scientific American. "Human, sea slug brains share genes for Alzheimer's and
Parkinson's." http://www.scientificamerican.com/article.cfm?id=human-sea-
slug-brains-sha.

Scientific American. "Strange but true: Whale waste is extremely valuable." http://
www.scientificamerican.com/article.cfm?id=strange-but-true-whale-waste-is-
valuable.

SeaWorld. "Penguins." http://www.seaworld.org/animal-info/info-books/penguin/
index.htm.

SeaWorld. "Walrus." http://www.seaworld.org/animal-info/info-books/walrus/index
.htm.

Shedd Aquarium. "Moon jellies." http://www.sheddaquarium.org/moonjellies.html.

Smithsonian Marine Station at Fort Pierce. "*Panulirus argus*: Spiny lobster." http://
www.sms.si.edu/IRLSpec/Panuli_argus.htm.

Smithsonian National Museum of Natural History. "Jellyfish romance—(*Carybdea
sivickisi*)." http://invertebrates.si.edu/jellyfish/showtime.html.

Smithsonian National Museum of Natural History, Department of Invertebrate Zoology. "World of copepods." http://invertebrates.si.edu/copepod/.

Texas A&M University. "Thinking big: Coccolithophores may be small, but they
know how to get attention." http://ocean.tamu.edu/Quarterdeck/QD5.2/shatto-
slowey.html#anchor102889.

TED Talks. "Sheila Patek clocks the fastest animals." http://www.ted.com/index
.php/talks/sheila_patek_clocks_the_fastest_animals.html.

Tree of Life. "Architeuthidae." http://tolweb.org/Architeuthis.

University of California, Berkeley. "Evolution of a biologist: Conversation with Roy
L. Caldwell." http://globetrotter.berkeley.edu/people/Caldwell/caldwell-cono
.html.

University of California, Irvine. "UCI biomechanics." http://biomechanics.bio.uci
.edu/_html/CV/cv_popular.html.

University of California Museum of Paleontology. "Introduction to Cubozoa: The
box jellies." http://www.ucmp.berkeley.edu/cnidaria/cubozoa.html.

University of California Museum of Paleontology. "Porifera: Life history and ecology." http://www.ucmp.berkeley.edu/porifera/poriferalh.html.

University of California Museum of Paleontology. "Secrets of the stomatopod."
http://www.ucmp.berkeley.edu/aquarius/.

University College London. Micropaleontology Division. "Foraminifera." http://www.ucl.ac.uk/GeolSci/micropal/foram.html.

University of Delaware. Voyage to the Deep. "Pompei worm." http://www.ceoe.udel.edu/deepsea/level-2/creature/worm.html.

University of Melbourne. Department of Biochemistry and Molecular Biology. "Cone shell." http://grimwade.biochem.unimelb.edu.au/cone/.

University of Miami. Rosenstiel School of Marine and Atmospheric Science. "National resource for *Aplysia*." http://aplysia. miami.edu/.

University of Michigan Museum of Zoology, Animal Diversity Web. "*Pycnopodia helianthoides*." http://animaldiversity.ummz.umich.edu/site/accounts/information/Pycnopodia_helianthoides.html.

University of New South Wales. "Blows reveal sex drive in whales." http://www.science.unsw.edu.au/news/blows-reveal-sex-drive-in-whales/.

University of Southern California. Caron Lab, Marine Environmental Biology. "The biology and ecology protists." http://www.usc.edu/dept/LAS/biosci/Caron_lab/index.html.

University of Tasmania. School of Zoology. "Guide to the marine zooplankton of southeastern Australia." http://www.tafi.org.au/zooplankton/imagekey/copepoda/index.html.

United States Antarctic Program. "The bloodless icefishes." http://antarcticsun.usap.gov/science/contenthandler.cfm?id=1540.

United States Environmental Protection Agency. Pesticides. "Chitosan: Poly-D-glucosamine (128930) fact sheet." http://www.epa.gov/pesticides/biopesticides/ingredients/factsheets/factsheet_128930.htm.

Walla Walla University. "*Pycnopodia helianthoides*" (sunflower star). http://www.wallawalla.edu/academics/departments/biology/rosario/inverts/Echinodermata/Class%20Asteroidea/Pycnopodia_helianthoides.html.

Wesleyan University. "Planktonic foraminifera and the ocean." http://ethomas.web.wesleyan.edu/ees123/forams.htm.

Woods Hole Oceanographic Institution. Dive and Discover. "Hydrothermal vents." http://www.divediscover.whoi.edu/vents/index.html.

Woods Hole Oceanographic Institution. Dive and Discover. "The watery world of salps." http://www.divediscover.whoi.edu/expedition10/hottopics/salps.html.

Woods Hole Oceanographic Institution. Oceanus. "The deepest divers." http://www.whoi.edu/oceanus/viewArticle.do?id=29067.

Woods Hole Oceanographic Institution. Oceanus. "Pilot whales—The 'cheetahs of the deep sea.'" http://www.whoi.edu/oceanus/viewArticle.do?id=41906.

World Asteroidea Database. http://www.marinespecies.org/asteroidea.

The Worldfish Center. "Reefbase: A global information system for coral reefs." http://www.reefbase.org.

World Resources Institute. "Reefs at risk—Analysis of threats to coral reefs." http://www.wri.org/project/reefs-at-risk.

Zea, S., T. P. Henkel, and J. R. Pawlik. "The sponge guide: A picture guide to Caribbean sponges." http://www.spongeguide.org.

ACKNOWLEDGMENTS

This book would not have been possible without the help of my colleagues, friends, family, and sponsors. At the heart of the book is the hard-won information that scientists have wrangled from the sea over years of study. Thank you for all that you do and share through publications in print and, now, online. I am especially grateful to those of you who have encouraged my efforts and provided specific material or photographs, or were willing to review portions of the text, including Rob Kramer and Drs. Allan Stoner, Mark Butler, Steven Miller, Christopher Mah, Ron O'Dor, and Roger Hanlon. Thanks also to Ann Campbell and the other librarians at the University of Miami's Rosenstiel School of Marine and Atmospheric Science, whose assistance to find even the most obscure journals and track down overdue books and whose love of the printed word were extraordinarily helpful and inspiring. My appreciation also goes to a pod of loyal and supportive friends, including Linda Glover, Doug and Melissa Ray, Robin Hawk, Lynn Martenstein, Cathy Sherrill, and Bob Wicklund. Their laughter is the stamp of approval on my continuing efforts to make science fun and engaging. Thanks also to my family for putting up with my distractions and waywardness, and their unfailing encouragement.

I especially want to thank the sponsors of this book. I have taken a very nontraditional track as a scientist, and financial support for my efforts is not always readily available. I was able to devote time and energy into the research and writing needed because of their generosity and our mutual goals in conservation, public education, sustainability, and in particular, an interest in and love for the oceans. Thanks goes to Walmart, the Cruise Lines International Association, and Celebrity *Xpedition* for their sponsorship as well as to The Ocean Foundation for acting as a creative works fiscal host to the project.

My sincerest appreciation also goes to copyeditor Carol Saller, whose keen eye and editorial skills helped to improve the text, and thank you for catching the errors that slipped by my word-weary eyes. And last, but in no way least, thanks to executive editor Christie Henry, book designer Matt Avery, and the others at the University of Chicago Press. As always your enthusiasm is an inspiration, and your support and advice invaluable.

SPONSORS AND PARTNERING ORGANIZATIONS

Walmart

Saving people money so they can live better was a goal Sam Walton envisioned when he opened the first Walmart store more than forty years ago. In 2005, Walmart made a commitment to become a more environmentally sustainable retailer. They adopted an approach called "Sustainability 360," which involves taking a comprehensive view of their business to find ways to reduce their own environmental impact while engaging suppliers, associates, and customers around the world in the company's sustainability efforts. Sustainability 360 recognizes Walmart's ability to make a difference on the environment, as well as the far greater results they can achieve by leveraging their entire global supply chain.

Cruise Lines International Association

Cruise Lines International Association (CLIA) is the world's largest cruise association, comprising the major cruise lines serving North America and nearly sixteen thousand travel agency members. CLIA exists to promote a safe, secure, and healthy cruise ship environment; to educate, train, and support its travel agent members; and to promote and explain the value, desirability, and affordability of the cruise vacation experience. Among the industry's many priorities, protecting the fragile environment in which they operate and implementing sound environmental practices are among the highest. The cruise line industry works with NGOs, universities, regulators, and scientists around the globe to continually improve its environmental practices. Visit www .CruiseIndustryFacts.com to learn more.

Celebrity Xpedition

Celebrity's ship *Xpedition* cruises year-round in the Galapagos Islands, and the program is committed to minimizing its impacts on the wildlife, water, and land of the region, while providing a first-class experience for its passengers and the best information available on site. The *Xpedition* program has won numerous awards for innovative environmental technology, their activities to reduce impacts in the Galapagos, and their commitment to helping the local community. Celebrity Cruises is part of Royal Caribbean Cruises, Ltd.

The Ocean Foundation

The Ocean Foundation is a unique community foundation with a specialized practice. Its niche is providing high-end philanthropic advice for a community of donors who care about the coast and oceans. TOF's mission is to support, strengthen, and promote those organizations dedicated to reversing the trend of destruction of ocean environments around the world.